T0185710

Lecture Notes on Data Engineering and Communications Technologies

Volume 50

Series Editor

Fatos Xhafa, Technical University of Catalonia, Barcelona, Spain

The aim of the book series is to present cutting edge engineering approaches to data technologies and communications. It will publish latest advances on the engineering task of building and deploying distributed, scalable and reliable data infrastructures and communication systems.

The series will have a prominent applied focus on data technologies and communications with aim to promote the bridging from fundamental research on data science and networking to data engineering and communications that lead to industry products, business knowledge and standardisation.

** Indexing: The books of this series are submitted to SCOPUS, ISI Proceedings, MetaPress, Springerlink and DBLP **

More information about this series at http://www.springer.com/series/15362

Jolta Kacani

A Data-Centric Approach to Breaking the FDI Trap Through Integration in Global Value Chains

A Case Study from Clothing Manufacturing Enterprises in Albania

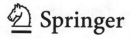 Springer

Jolta Kacani
Faculty of Economics
University of Tirana
Tirana, Albania

ISSN 2367-4512   ISSN 2367-4520   (electronic)
Lecture Notes on Data Engineering and Communications Technologies
ISBN 978-3-030-43188-4   ISBN 978-3-030-43189-1   (eBook)
https://doi.org/10.1007/978-3-030-43189-1

This Springer imprint is published by the registered company Springer Nature Switzerland AG
The registered company address is: Gewerbestrasse 11, 6330 Cham, Switzerland

To my mum Violeta and to my father Jorgaq.

To my only brother Ardit.

Preface

Emerging economies experience significant regional disparities that bring an ongoing challenge to strengthen regional development and to achieve economic convergence with developed economies. One-way emerging economies can obtain regional convergence and achieve sustainable growth is to take out the benefits from the presence of foreign direct investment (FDI) as an instrument to promote integration into global value chains (GVCs). With reference to industrialization induced by FDI and with special focus in the clothing manufacturing industry, this book proposes a data-centric approach to look into whether the FDI trap of industrialization is broken. This approach is based on time series analysis at a country and industry level and on data sets retrieved from international databases such as EORA, ICIO, and World Integrated Solutions.

The data-centric approach is directly linked to the knowledge transferred in the host territory and the evolution in the quality of clothing manufacturing enterprises. The validity of the framework is obtained from a case study methodology and a profound analysis based on data-driven decision-making and organizational commitment to obtain integration in GVCs. Moreover, for each area of the proposed framework this book introduces policy actions to enhance a virtuous FDI circle that leads to upgrading at the enterprise, industry, and country level. The desired upgrading enables emerging economies to become active members in existing GVCs including countries in the Western Balkan region.

This book presents challenges and opportunities for emerging economies, and it embraces different aspects of integration in GVCs obtained from statistical analysis of time series data sets. This book presents different methods with regard to data collection, operational activity, business analytics, and mobility of enterprises within GVCs in the clothing industry. This book highlights the need for better innovation policies in developing countries as a precondition for active membership into GVCs. The casual effects resulting from robust statistical models indicate that innovation in exports improves trade openness and turns emerging economies into active players of GVCs. With reference to best fitted statistical models, emerging economies need to upgrade their degree of innovation based on the knowledge transferred and linkages established from the presence of FDI. Finally, this book presents a detailed case study methodology based on data collection in the Western Balkan countries with special focus in Albania and it introduces strategies for enterprise upgrading and increased absorptive capacity.

I decided to write this book in order to increase global awareness on the social and economic developments occurring in emerging economies, as they can become the new industrial hubs in existing GVCs. This book is also a step forward in the limited research undertaken so far on the Western Balkan countries and it also aims to advocate the high potential of the region as the new industrialization and innovation region of the European Union. The Western Balkan region can turn into the most beneficial manufacturing outsourcing hub for the European Union, while the market can retain its global competitiveness at times when it is challenges the most from fully established GVCs in Asia and beyond.

Tirana, Albania   Jolta Kacani

Acknowledgements

This book would not have been possible without the guidance and support of many people. To all of them, I owe my deepest gratitude and immense appreciation.

I deeply appreciate the precious guidance, advice, feedback, and recommendations of Prof. Fatos Xhafa of Universitat Politècnica de Catalunya that contributed to realization of this book. I will always be grateful for his unconditional support.

I would like to thank you Gjergji Gjika, President of the Chamber of Façon in Albania, for his helpful insights on the clothing industry and his participation in data collection during fieldwork.

In particular, I would like to express my gratitude to Prof. Ricardo Hausmann and the team from Harvard Kennedy School of Government. From this outstanding team, I had the privilege to acquire in-depth knowledge on how to implement the first industrial policy in the clothing manufacturing industry.

Special thanks to the management and employees in the four clothing manufacturing subsidiaries. I am grateful for their time, hospitality, and detailed explanations provided on the operational and manufacturing activity. More specifically, to Gjergj Leqejza and Ema Thani in Shqiperia Trikot sh.p.k; Bernd Naber and Ornela Koxhaj in Naber Konfeksion sh.p.k; Dario Albeni and Selim Dyrmishi in Valcuvia Alba sh.p.k; and Christos Diamantidis, Edmond Haxhi, and Mariana Raidhi in Industria Ballkanike sh.p.k.

Of great importance was the support of colleagues and friends in national institutions that made accessible numerical data on FDI and the clothing industry. More specifically, I am grateful to Elsa Dhuli, Etugert Llaze, Alma Mara, and Pranvera Elezi from INSTAT; Endrit Lami from the Ministry of Finance; Diana Shtylla and Iris Mele from the Bank of Albania; and Albana Zykaj, Ada Elezi, and Dorela Ceka from AIDA.

I am grateful to Prof. Marie Löwegren in Sten K. Johnson Centre for Entrepreneurship, Lund University School of Economics and Management, for providing new insights and approaches on innovative ecosystems as a prerequisite for facilitating regional integration among enterprises in the Western Balkan region. Special thanks to Andreas Bryngelson in Lund University Commissioned

Education for showing me the impact of science parks to generate innovation in emerging economies.

My deepest appreciation to Prof. Annie Triantafillou, Prof. Anna Giannopoulou, and Prof. Dimitris Doulos at the American College of Greece for inspiring and motivating me throughout my challenging years as an undergraduate student. I will always be grateful for their unconditional support.

A special appreciation to Prof. Massimo Guidolin at Bocconi University and the University of Manchester for opening new dimensions on conducting research for motivating and for believing in me.

A very special attribute to Jorgjeta Marko, Iris Fekollari, Arlind Asllani, Silvi Peta, and Anisa Isufi for making such an amazing team to work with on data collection and analysis.

I am much thankful to Jennifer Sweety Johnson, Raghavy Krishnan, Thomas Ditzinger, Anja Seibold, and Holger Schaepe etc. for their precious support and guidance in publishing this book.

Finally, I am much thankful to my precious family. Their motivation, patience, and immense support have been priceless in preparing this book.

About This Book

This book provides insights on how emerging economies can turn into success stories of sustainable development through integration in GVCs. The data-centric approach followed in this book is based on data analysis and case study methodology. Time series analysis is implemented at a country and industry level on data sets obtained from international donors like the World Bank and United Nations Conference on Trade and Development (UNCTAD) such as EORA, ICIO, and World Integrated Solutions.

The causal effects of the current innovation on the trade performance aim to contribute to the existing literature on the effect of innovation in improving export performance in emerging economies in general and in the Western Balkan region in particular. With reference to best fitted statistical models, emerging economies upgrade their degree of innovation based on the knowledge transferred and linkages established from the presence of FDI.

In addition, the data-centric approach of this book is supplemented with a case study methodology. The framework serves as the reference point for analyzing both the quantitative and the qualitative effects of foreign clothing manufacturing enterprises that intend to become trusted suppliers of lead firms that dominate operations in GVCs. The proposed framework consists in the knowledge transferred in the host territory and the evolution in the quality of the subsidiary over time.

Chapter 1 presents the rationale, the objectives, and the data-centric approach followed in this book. Chapter 2 is dedicated to a thorough literature review focused on the qualitative and quantitative effects of FDI in the host territory. This chapter presents also the evolution in the activity of the subsidiary, mobility potential, contrasting views on the impact of FDI in a host territory, and concludes with examples of policies implemented for attracting FDI in various emerging economies. Chapter 3 is dedicated to the advantages developing countries obtain from participation in GVCs. This chapter brings statistical models on the current integration of emerging economies in GVCs and identifies key determinants for active participation. Chapter 4 presents the clothing industry. The first part of the chapter is dedicated to the recent trends in the clothing industry including activities in GVCs, product categorization, industry upgrading, type of operating firms, and the trend of fast fashion. The second part of the chapter refers to the clothing industry in Albania with special focus on data analysis to identify the existing integration of this industry in GVCs. Chapter 5 presents the case study methodology of the data-centric framework including its advantages and limitations, examples of similar methodology in the literature, steps followed in analyzing the four case studies, types of validity, and generalization in case study methodology. Chapter 6 introduces a data-centric description of the four clothing manufacturing enterprises that aim to become trusted suppliers of lead firms that dominate operations in GVCs. Chapter 7 presents a data-centric comparison of the case studies and compares them with reference to previously introduced framework. Chapter 8 highlights the importance of innovation as an instrument for sustainable development and uses statistical models to estimate the impact of innovation with regard to trade openness in emerging economies. Chapter 9 draws conclusions, provides policy recommendations, and suggests areas for further research for an active participation of emerging economies in GVCs.

Contents

About the Author

Jolta Kacani Ph.D. is currently a Manager in the Consulting Department in Deloitte Central Europe, where she has been working for more than five years. Her responsibilities include building partnerships with major international enterprises based on industry specifics and managing awarded projects financed from international organizations like the World Bank, European Commission, United Nations, European Bank for Reconstruction and Development, and the German Government. She has an extensive experience in the management of complex donor-funded projects with high impact on the economic development of the Western Balkan region including Albania. She is a certified international professional in project management from the International Labour Organization (ILO).

She was previously the Director of Innovation, Technology and Statistics in the Ministry of Economy, Trade and Entrepreneurship of Albania, where she was responsible for identifying, preparing, and implementing strategic projects related to innovation and technology. She was also Head of European Integration and Strategic Projects Sector, Head of National Quality Infrastructure Sector, and Head of Development of Economic Zones Sector at the Ministry of Economy, Trade and Energy.

For over twelve years, she has pursued an intensive academic career by being Lecturer and Researcher for modules such as International Corporate Finance, Research Methods in Finance and Accounting, International Markets and Institution in various universities in Europe and Albania.

In addition, she is Advanced Licensed Trainer and Coach from CEFE International in Germany. She has trained more than 500 participants, in developing business skills, and has coached more than 75 participants in successfully presenting their business ideas and in drafting sustainable business plan.

She holds a Ph.D. in international economics from Universitat Politècnica de Catalunya—BarcelonaTech, M.Sc. in Finance and Economics from the University of Manchester, and a B.A. in Economics from the American College of Greece.

She holds also a certificate from the Sten K. Johnson Centre for Entrepreneurship Summer Academy for Young Professionals, on Innovation Ecosystems and Entrepreneurship, obtained at Lund University, Sweden, and holds a number of professional certifications from international organizations.

Abbreviations

ADA	Austrian Development Agency
AIDA	Albanian Investment and Development Agency
ATC	Agreement on Textile and Clothing
BoA	Bank of Albania
CMT	Cut, Make, Trim
DCM	Decision of Council of Minister
ECR	Efficient Consumer Response
EDI	Electronic Data Interchange
EPZ	Export Processing Zone
EU	European Union
FDI	Foreign Direct Investment
GDP	Gross Domestic Product
GERD	Gross Expenditure on Research and Development
GII	Global Innovation Index
GIZ	Deutsche Gesellschaft fur Internationale Zusammenarbeit
GoA	Government of Albania
GOTS	Global Organic Textile Standard
GVCs	Global Value Chains
ICT	Information and Communication Technology
ILO	International Labour Organization
INSTAT	National Statistical Institute of Albania
MFA	Multi Fiber Arrangement
MNE	Multinational Enterprise
MoF	Ministry of Finance
MPS	Modular Production System
NPL	Non-Performing Loan
NRC	National Registration Center
OBM	Original Brand Manufacturing
OECD	Organisation of Economic Co-operation and Development
OEM	Original Equipment Manufacturing

OLS	Ordinary Least Squares
PBS	Progressive Bundle System
POS	Point of Sale
QRS	Quick Response System
R&D	Research and Development
RESET	Ramsey Regression Equation Specification Error Test
SME	Small and Medium Size Enterprise
UNCTAD	United Nations Conference on Trade and Development
UNDP	United Nations Development Programme
UPS	Unit Production System
USAID	US Agency for International Development
WB	Western Balkans

List of Figures

List of Tables

Chapter 1
Introduction

Abstract This chapter sets the background on the importance participation in global value chains has for emerging economies with regard to industrialization and sustainable development. A way for emerging economies to achieve sustainable development is to reap the benefits of foreign direct investments by breaking the middle-income trap. The chapter also presents the data centric approach based on time series analysis and the case study methodology with special focus in the clothing manufacturing industry.

Keywords Data centric approach · Virtuous FDI circle · Vicious FDI circle · Global value chains · Sustainable development

1.1 The Importance of FDI Induced Industrialization for Participation in Global Value Chains

In the last two decades, the structural changes occurring in the global economy have remodeled global production and trade causing changes in the organization networks of various industries and national economies. The geographical fragmentation of industries and the division of product value in different countries before it reaches the end consumer has resulted in upgrading of processes along global value chains (GVCs) [1]. The increasing fragmentation of production facilitates the connection of geographically dispersed production units into a single industry and serves to identify the ever-changing patterns of trade and production. On a macroeconomic level fragmentation in GVCs helps to understand the interaction the interconnection among national economies. The fragmentation of production has encouraged economic growth in emerging economies leading to an increase in the global demand for goods and services as well as increase in international trade transactions [2].

The fragmentation of production permits emerging economies to participate in GVCs. The presence of foreign direct investment not only facilitates integration but also turns emerging economies into active players of GVCs. In order for this to happen emerging economies need to break the vicious circle of FDI and encourage a virtuous circle (see Fig. 1.1). A vicious circle occurs when the host territory attracts low quality FDI with principal motivations to benefit from the cheap labor force and access to

© Springer Nature Switzerland AG 2020
J. Kacani, *A Data-Centric Approach to Breaking the FDI Trap Through Integration in Global Value Chains*, Lecture Notes on Data Engineering and Communications Technologies 50, https://doi.org/10.1007/978-3-030-43189-1_1

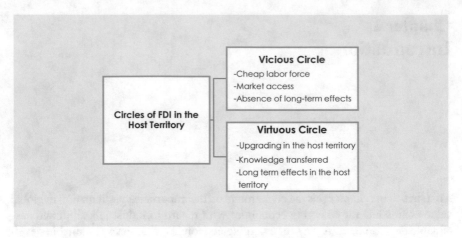

Fig. 1.1 FDI induced circles in the host territory. *Source* Author's drawing based on Dicken [3]

markets. This circle is very unstable and generates limited knowledge transfer in the host territory. These motivations do not improve location advantages but expose the host territory to a FDI trap without being able to maximize the benefits from the presence of foreign investors. The cheap labor force with lower wages compared to neighboring countries and the rising costs encountered in Asian countries, are temporary location advantages that appeal to foreign investors to transfer production in emerging economies. However, when the temporary location advantages are no longer beneficial to foreign investors production is transferred to other locations leaving the existing host territory without integration in GVCs.

For emerging economies breaking the FDI trap means to come out of the middle-income trap and move into higher stages of economic development, including creativity and innovation (see Fig. 1.2). In order to break the middle-income trap the host territory needs to acquire additional knowledge, to establish developmental linkages with FDI, and to upgrade production activities [5]. By focusing on these effects, this research tries to identify the benefits of the host territory occurring from the FDI production activity in case they decide to move out and transfer production to another host territory.

The findings of this book are generated based on a data centric approach at a country and industry level and based on time series obtained from databases of international donors like the World Bank, the United Nations Conference on Trade and Development (UNCTAD), and the European Commission. With reference to industrialization induced by FDI and with special focus in the clothing manufacturing industry this book proposes a data centric approach to look into whether the FDI trap of industrialization is broken. This approach is based on time series analysis at a country and industry level and on data sets retrieved from international databases such as EORA, ICIO, and World Integrated Solutions.

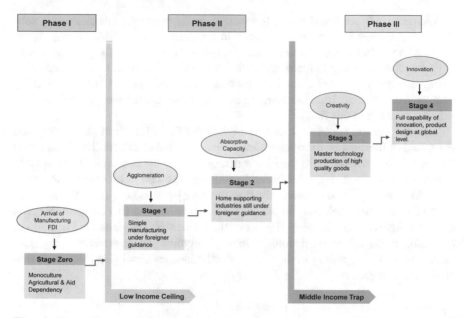

Fig. 1.2 Phases of economic development in emerging economies. *Source* Author's drawing based on Ohno [5]

The clothing industry it is often regarded as an employment generator with regard to economic development and as the first step of industrialization for emerging economies. Following the results obtained from data analysis, a further step of the data centric approach included in this book is a framework to analyze the impact of FDI in the clothing industry as an instrument to strengthen integration of emerging economies into GVCs. The framework is designed based on a data centric case methodology as it offers a number of advantages including direct observation in the unit of analysis during fieldwork, consideration of many quantitative and qualitative variables, flexibility, and feedback obtained through interaction with stakeholders relevant to the area of interest to the research. Four clothing manufacturing subsidiaries were selected as case studies for this research. Two of them have Italian ownership, one German, and one Greek [4] (Fig. 1.3).

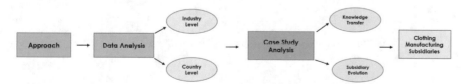

Fig. 1.3 The data centric approach based on data analysis and case study methodology. *Source* Author's drawing

The case study methodology undertaken in the four clothing manufacturing subsidiaries consisted in intensive fieldwork in the production facilities and in interviews with the management and assembly employees. The case study analysis goes in depth of the clothing industry by getting the whole picture of process occurring during production in the clothing industry and by analyzing the evolution of clothing manufacturing subsidiaries in emerging economies despite of the limited data and existing research in this field.

The data centric approach looks into whether FDI in the clothing industry has generated in the host territory quantitative and qualitative effects that include: (i) the knowledge transferred in the host territory and (ii) the evolution in the quality of clothing manufacturing subsidiaries in Albania. The first pillar of the framework refers to knowledge transferred in the host territory by considering: (i) the group to which the subsidiary belongs, (ii) the stock of knowledge, and (iii) the channels used to convey knowledge. The second pillar of the framework concerns the evolution in the quality of the subsidiary by looking into: (i) upgrading in the sense of complexity of activities, (ii) upgrading in the sense of embeddedness, and (iii) the factors that affect upgrading.

References

1. Belderbos R, Sleuwaegen L, Somers D, De Backer K (2016) Where to locate innovative activities in global value chains: Does co-location matter? OECD science, technology and industry policy papers 30. OECD Publishing, Paris
2. Borin A, Mancini M (2019) Measuring what matters in global value chains and value-added trade. Policy research working paper no. WPS 8804, WDR 2020 background paper. World Bank Group, Washington DC
3. Dicken P (2011) Global shift: mapping the changing contours of the world economy, 6th edn. The Guilford Press, New York
4. Kacani J, Van Wunnik L (2017) Using upgrading strategy and analytics to provide agility to clothing manufacturing subsidiaries: with a case study. Glob J Flex Syst Manag 18(1):21–31
5. Ohno K (2006) The economic development of Japan. GRIPS Development Forum, National Graduate Institute for Policy Studies, Japan

Chapter 2
FDI as an Instrument for Sustainable Economic Development of the Host Territory

Abstract This chapter highlights on the impact the presence of foreign direct investments has in the sustainable development of emerging economies. It starts with introducing the OLI paradigm and technology as a determinant for economic development followed by the quantitative and qualitative effects emerging from the activity of foreign direct investment in the host territory. Additional areas covered in the chapter include the evolution of the subsidiary from upgrading in the operations and activities undertaken in the host territory together from the quality and the intensity of linkages established with local suppliers. The chapter continues with two opposing views on the impact foreign direct investment has in the developing path of the host territory categorized as the optimists and the pessimists. The chapter concludes with policies to attract and to foster upgrading in the manufacturing industry applied in various emerging economies.

Keywords OLI paradigm · Knowledge transfer · Pool of suppliers · Subsidiary upgrading · Developmental linkages

2.1 Introduction

Throughout history, nations have followed different paths to obtain prosperity. Some of the paths followed turned into success stories while others produced undesirable outcomes. Still today, policy makers and economic agents have yet to discover the path that would lead to sustained prosperity. Discovering the right path is the first step while picking up the right combination of instruments to walk along is the other step. The scope of this chapter is to introduce theories of economic development and the effects of FDI in the development of host territories. In this research, reference to FDI is made based on the approach of International Monetary Fund (IMF)[1] as "the net inflows of investment to acquire a lasting management interest (10% or more of voting stock) in an enterprise operating in an economy other than that of the investor".

---

[1] Please refer to www.imf.org.

© Springer Nature Switzerland AG 2020

J. Kacani, *A Data-Centric Approach to Breaking the FDI Trap Through Integration in Global Value Chains*, Lecture Notes on Data Engineering and Communications Technologies 50, https://doi.org/10.1007/978-3-030-43189-1_2

This chapter starts with introducing the OLI paradigm and technology as determinants for economic development. It continues with the economic impact of FDI in the host territory including quantitative and qualitative effects. Additional areas covered in chapter include the evolution of the subsidiary and the two opposing schools on FDI impact on the host territory categorized as the optimists and the pessimists. The chapter concludes with policies to attract and to foster upgrading of inward FDI applied in various emerging economies.

2.2   The OLI Paradigm Answering What, Where and Why?

The eclectic paradigm introduced by Dunning [33] is a well-known framework of international production. The OLI paradigm helps to understand better the motives to locate and maintain production in the host territory. The paradigm seeks to identify groups of relevant variables to explain foreign production. The eclectic paradigm sets three conditions that influence the decision of MNEs to place production in a host territory (see Fig. 2.1). Reference is made to Organization of Economic Co-operation and Development (OECD)[2] that regard multinational enterprise (MNE) as a set of companies or other entities established in more than one country that are connected and may co-ordinate their operations in various ways. While one or more of these entities may be able to exercise a significant influence over the activities of others, their degree of autonomy within the enterprise may vary widely from one MNE to another.

The first condition addresses the question of what characteristics firms going abroad possess that other firms do not have. This condition relates to competitive advantages successful MNEs own which allow them to compete with other firms

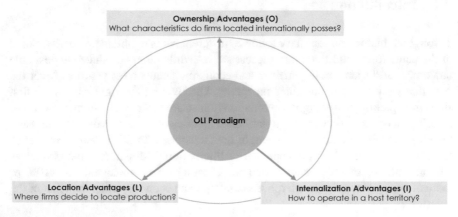

Fig. 2.1 The OLI paradigm. *Source* Author's drawing

---

[2]Please refer to www.oecd.org.

and to overcome the costs of operating in host territories. Dunning [34], refers to these competitive advantages as ownership advantages (O). These advantages refer to characteristics (patents, production techniques, entrepreneurial skills, modern technology, access to markets) a firm should possess to expand its activity in new host territories. They are also known as competitive advantages. Ownership specific advantages include also innovative capabilities, capable managers, and advanced technology [64]. These ownership advantages give to firms possessing them a competitive advantage over their competitors in settling production in various host territories [34].

The second condition addresses the question where MNEs choose to locate their production. This condition relates to country specific advantages that MNEs find attractive for locating production activities. These are called location advantages (L). Location (L) are advantages that a host territory should possess in order to be able to attract MNEs. They can be categorized into physical advantages (strategic geographical position, proximity to markets, natural resources, etc.) and business environment resources (market access, labor costs, qualified and specialized labor, base of competitive and specialized suppliers, special tax incentives, etc.) [20]. The paradigm states that it is more profitable for firms to utilize ownership advantages in conjunction with the advantages available in the host territory; otherwise, MNEs will just continue to produce in the home territory and export production to the host territory. Today, location advantages are oriented toward opportunities for MNEs to create in cooperation with local firms assets that will enhance their competitive advantages [35]. The third condition addresses the question on how MNEs decide to operate in a host territory. MNEs may operate with their own production units or they may subcontract production to other firms. MNEs choose to set up their production units in order to fully utilize competitive advantages and to minimize transaction costs. Dunning [35] names them internationalization advantages (I). This is why the eclectic paradigm is also referred to as the OLI paradigm [112]. Internalization advantages (I) relate to the ability of MNEs to make the best use of its competitive advantages. Dunning [34] explains that MNEs need to possess an internalization advantage in order to exploit their competitive advantage internally rather than offering to other firms through some contractual arrangement such as subcontracting or licensing. In this case, MNEs prefer to transfer their competitive advantages across national boundaries within their own organization rather than subcontract them [56, 111]. MNEs can establish an internal market within their administrative organization to avoid high and uncertain transaction cost associated with asymmetric information (the seller has more information than the buyer) and the principal agent problem (a contractor of MNEs behaves on its own interest and not doing what the MNE requires) [34, 35].

Despite of its advantages and wide citation in the international literature, researchers criticize the OLI paradigm. Part of this criticism is also present in the papers written by Dunning [34, 35] himself. The eclectic paradigm is considered as a process with little guidance on how MNEs organize their operational activities in order to acquire and generate advantages that are more competitive in the future [60].

Also, the paradigm concentrates on profit maximization activities of MNEs overcoming any functions and operations they may perform apart from profit maximization like research and development, social responsibility, etc. [35].

Moreover, scholars argue that O, L, I advantages rely on numerous variables will little predictive value on day-to-day operations of MNEs. It is very difficult to judge based on the explanatory variables of the paradigm for how long MNEs will continue to operate and produce in a host territory [24, 82].

2.3   Knowledge: A Key Determinant for Sustainable Economic Development

This section introduces views on theories of economic development based on sustainable economic growth and the factors needed o obtain such growth in emerging economies. In this research, the economic development is based on the views of World Bank[3] as qualitative change and restructuring in a country's economy in connection with technological and social progress reflecting an increase in the economic productivity and average material wellbeing in the population of a given country. Economic development is closely linked to economic growth. However, in order for territories to develop they need to obtain economic growth that is sustainable. This is achieved when economic resources like stock of physical capital, stock of human capital, stock of natural resources, and knowledge increase and/or are used more efficiently [8].

According to Hausmann and Rodrik [51] and Newman et al. [86], sustainable growth requires not only use of knowledge but also institutions that are able to achieve such a growth. This perspective is well grounded in the neoclassical model of economic growth, which predicts that poor countries will experience rapid convergence with advanced economies once they have access to state-of-the-art knowledge and their governments respect property rights [65, 76].

Any process with regard to economic development depends to a certain extend on the quality of institutions and good governance. Institutions are the ones that respond to the needs of economic and social actors. Institutional response can either facilitate economic activity or obstruct it [31]. Increased competition in international markets demands efficient responses and strategic cooperation of institutions. Many institutions have undertaken effective strategies to respond to the needs of economic, social, and political dynamics [9, 106].

In the largest and most innovative regions, ties among institutions have become complex while the number of institutions has multiplied. Amin and Thrift [6] refer to this phenomenon as "institutional thickness". Despite of the presence of institutional thickness, in regions where institutions are flexible economic development is the strongest [63]. Flexible institutions provide room for lowering transaction costs, building trust among economic and social agents, and forming of strategic alliances.

---

[3]Please refer to www.worldbank.org.

Institutions become a factor in the process of capital accumulation and collective learning to the extent to achieve economies of scale and scope [2, 101].

Referring to Vasquez-Barquero [114], local economic development agencies are among the main institutions having a significant role in achieving prosperity in the regions. These are non-profit organizations operating under a private and public capital and have as their key objectives: (i) to create and develop a friendly business environment necessary for enterprises to start up, (ii) to provide support services for social involvement of regions, (iii) to establish strong connections among economic agents, (iv) and to increase productivity and innovation capabilities of local enterprises [73, 108].

Even though at the core of development are local institutions, the central administration has also an important role in supporting regional development technically and financially. Kowalski et al. [66] advocates that state of the art entrepreneurship activities have initiated from proactive government interventions. International organizations have supporting programs and promoting strategies that can make regions eligible for central government financing. Rodrik [95] argues that central institutions are stronger and larger in the most developed countries where regions are more efficient in generating wealth once they are supported by reliable central institutions [120]. Sustainable growth redirects attention away from capital accumulation, plant capacity, and acquisition of equipment to immaterial resources like innovation, human capital, and knowledge [66, 70].

With reference to Rodrik [98, 97], the main determinants for sustainable growth are the stock of physical and human capital together with knowledge, and innovations. For Easterly [36], technology is the organized knowledge that coordinates workers and machines while Freeman [41], argues that technology is a bulk of knowledge about techniques Changes in knowledge lead to innovation. According to Dicken [27], innovation is inventing and spreading of new methods of doing things. For Easterly [36], technological change mean the blueprints are improved. Changes in technology lead to higher productivity. Technological change is the only way in which output per worker (i.e. productivity) can keep increasing in the long run. Acquiring knowledge is nowadays a challenge and an objective of many countries especially in emerging economies.

To continue, the change in productivity depends on how labor merges with other components of production, on how efficiently equipment is used, and on how new knowledge is incorporated into operational activities [99]. Economists highlight the role of the government, of the private sector, of institutions, and of markets in fostering innovation and stimulating workers to be more productive [1, 103].

Knowledge is widely recognized as a source of economic development as it stimulates economic growth and increases flexibility in regional structures of production systems. Martinus and Sigler [78] together with Easterly [36] argue that technology is of outmost interest to countries as it belongs to the core of economic development.

Introduction and diffusion of new knowledge alters the dynamics of production systems. In the process of collective learning, coded and tacit knowledge are diffused in the region through alliances established in the industrial network in which knowledge is spread. Even though this process is conditional on the characteristics of the region, experience, and history of the local market economy, knowledge has a vital role in spreading innovation [107].

Krugman and Obstfeld [69], advocate that knowledge is becoming along with labor and capital a main factor in production. Acquiring new knowledge is fundamental for economic catch up and sustained growth in regions. Romer [100], emphasizes that knowledge ranges from abstract ideas such as scientific formulae to a practical applications such as a traffic circle.

Moreover, innovation is a collective learning process in which enterprises make investments and location decisions [30]. Internalization of innovation permits enterprises to expand their product range, to create more economically efficient plants, and to strengthen economies of scale. Innovation creates opportunities to undertake strategies for entering new markets and for enhancing competitiveness by introducing improved products [29, 67].

Many countries have launched various policies to support and fasten diffusion of knowledge, spread innovation, and adapt new technologies through the establishment of innovation centers, scientific parks, and technological institutes. For example, in Brazil, technological centers to support shoe producers to reach and maintain the quality of standards required for export in overseas markets were established [39, 40].

In the 1980s government, policies in Japan focused on supporting disadvantaged regions through programs for obtaining technological progress in information and communication (ICT) industries [11]. In Malaysia, the initiative to establish industrial parks as instruments to facilitate diffusion of knowledge and innovation was strongly supported. These parks provide financial support to individuals that aim to turn an innovative idea into a business venture, assist enterprises in implementing research projects, and to offer training services for enterprises operating in electronics and information industries [92, 93].

2.4 Economic Impact of FDI in the Host Territory

FDI impact in host territories has captured the attention of scholars, governments, and citizens. It has generated controversial views on how they affect economic development and long-term sustainable growth. The impact of FDI in host territories is broadly categorized into: (i) quantitative effects and (ii) qualitative effects. Quantitative effects relate to the impact of FDI on macroeconomic indicators of host territory and they are short-run. Qualitative effects are linked to technological and pecuniary external economies resulting in a host territory from the presence of FDI activity. These effects are long-term in nature. Generalizing on the impact FDI has in the economic development of the host territory is not possible as it depends on: (i) the characteristics of FDI and (ii) the characteristics of the host territory.

2.4.1 Quantitative Effects (Short-Run) Effects

This section presents the quantitative effects of FDI in host territories. Main effects include job creation, contribution to GDP, fluctuations in the physical capital stock, and changes in tax revenues of the host government.

Job creation is one of the main quantitative effects of FDI in host territories. Employment generated through the presence of FDI is both direct and indirect. Direct employment is generated as a need for a subsidiary to perform its activities. It depends mainly on the type of the industry and on the scale of operations [84]. Subsidiaries operating in labor intensive industries (textile, shoe making) create a high number of jobs primarily blue collar ones while those operating in capital intensive industries create fewer jobs mostly white collar ones. Indirect employment depends on the linkages a subsidiary establishes with the local economy. Subsidiaries that purchase raw materials, intermediate goods, or services in host territories bring more revenues to local suppliers that in turn have better opportunities to make investments and to increase the production activity and to generate employment [21]. As employment among local suppliers goes up so does consumption. Higher consumption generates more employment in various sectors of the economy, leading to a multiplier effect in a host territory [4, 28, 79].

On the other hand, Colen et al. [23] and Gachunga [43] provide a different perspective on FDI created employment. They argue that the presence of foreign firms increases competition in host countries, which results in crowding out of local firms. In order to make a judgement on the crowding out two effects need to be taken into account. The first effect concerns competition on the market, whether the production of the subsidiary is for the domestic market of the host territory or it is exported. Competitive advantages allow MNEs to obtain lower production costs, to shift domestic demand towards their goods and services, and to attract skilled labor [49]. The second effect occurs in the labor market as foreign firms attract skilled labor from local firms, leaving them with the less skilled ones. This occurs because local firms cannot pay the same wage as the foreign investor that usually pays higher. The impact on employment will depend on the net effect as they may experience a decrease in demand and are unable to hire skilled labor [55].

In addition to employment, FDI activity contributes to the GDP of the host territory. The contribution can be direct and indirect. Direct contribution to GDP is assessed by the production and expenditure approach [37]. The production approach is linked to value added. Value added is total sales less the value of intermediate inputs used in production. On the other hand, the expenditure approach refers to the sum of consumption (C), investment (I), government expenditure (G), and the difference between imports and exports (X − M).

FDI can affect GDP components of the expenditure approach positively and negatively. Indirectly, FDI can increase consumption in the host economy through an increase in employment and provision of wages higher than those offered locally can. Consumption in the host territory can also go down as FDI may crowd out domestic firms and may lower employment in the host economy [26]. In addition, in this case the net effect will depend on whether production is for the local market or exported and on the employment of skilled labor previously working in local firms.

To continue, the mode of entry of FDI is a determinant of the level of investment in the host territory. Greenfield FDI accompanied with establishment of new production facilities lifts the level of investment in the host territory. FDI in the form of mergers and acquisition does not affect the level of investment [27]. Moreover, the level of exports and imports in a host territory is affected by FDI activity. The policy orientation of the MNE on the destination of final products determines the level of exports it generates. The level of exports goes up if final output is designated for foreign markets. A similar scenario happens with imports. When inputs required in production of final goods are not available in the host territory in either quality or quantity the purchase of inputs abroad goes up resulting in an increase in the level of imports [5]. Nevertheless, the effect of the difference between exports and imports on GDP is uncertain as it depends on whether the destination of production of the subsidiary is the domestic or the export market [60, 75].

Another quantitative effect of FDI is the change it causes on the level of capital stock. According to Nehru and Dhareshwar [83] physical capital stock refers to goods that are fixed, tangible, durable, and reproducible that are necessary to realize production. These investments may cause an increase in the stock of physical capital depending on how FDI enters in the host territory. When considering how FDI affects the stock of physical capital a distinction should be made between greenfield investments and mergers and acquisitions. Greenfield investments increase the stock of physical capital as operating facilities are raised from the ground [20]. On the other hand, mergers and acquisitions have a questionable impact on the stock of physical capital. Mergers and acquisitions may reduce the stock of physical capital as they try to achieve economies of scale by removing duplicate departments or closing down inefficient production plants. However, MNEs may undertake additional investments especially in modern technology when they acquire or merge with a firm having a high growth potential, leading to an increase in the stock of physical capital [18, 65].

FDI may bring additional revenues to governments in host territories if they are subject to taxation. As host territories attract more FDI, the number of entrepreneurship entities operating in the local market goes up [63]. An increase in entrepreneurship activities that is subject to taxation, translates into more tax revenues for host governments. Governments can use tax revenues to increase spending on public goods or to finance pending reforms [56]. The net effect will depend on tax revenues from employment created versus unemployment created from the presence of foreign firms and fiscal incentives provided to foreign firms from operating in a host territory. For example, in some African countries tax revenues coming from FDI are a vital source of the government budget. For example, in Botswana, 50% of government revenues in 2007 came from taxation paid by MNEs operating in the mining industry [23].

2.4.2 Qualitative Effects (Long-Run) Effects

Quantitative effects are short-run effects. What is more rewarding for developing countries are long-run effects occurring from the presence of FDI that bring about structural change, i.e. knowledge transfer and restructuring of the host economy. This section focuses on the qualitative effects of FDI in host territories. These effects fall into: (i) technological externalities and pecuniary externalities, (ii) channels of knowledge transferred in the host territory, (iii) the local absorptive capacity, (iv) and the presence of a pool of specialized labor and suppliers in the host territory [7].

Scitovsky [104] differentiates two categories of external economies. He refers to the first category as technological external economies while he calls the second category as pecuniary external economies (see Fig. 2.2). According to Krugman [68], external economies increase the strength of industry in one location. He argues that technological external economies occur when knowledge flows among firms and they learn from each other increasing the competitive advantages of the industry located in the host territory. Pecuniary external economies depend on the size of the

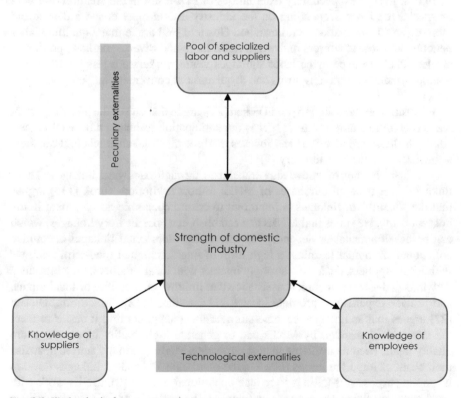

Fig. 2.2 Technological versus pecuniary externalities from the presence of FDI. *Source* Author's drawing based on Krugman and Obstfeld [69]

market. A strong domestic industry offers a large market for specialized labor and suppliers, which in turn make the industry even stronger [68, 49].

Moreover, MNEs create in host territories richness and density of local firms creating a pool of specialized local suppliers. As more MNEs start to operate in host territories, more specialized local suppliers can appear, as there is an increase in the demand for specialized goods and services [27, 78]. Size is also a factor in the creation of specialized suppliers. More specialized suppliers are often present in large industries rather than small ones [69]. The pool of specialized suppliers is created in two ways. Firstly, as MNEs demand more inputs they open up new markets for the provision of goods and services that suppliers can exploit as new opportunities. For example, if a firm grows larger and increases its orders from suppliers, they may be able to enjoy lower costs through economies of scale in their production. Secondly, MNEs in host territories "force" existing suppliers to compete with each other. Krugman and Obstfeld [69] argue that with a pool of specialized suppliers key inputs are less costly and more easily available as there are many firms trying to provide them. As the pool strengthens, suppliers concentrate on what they can do best subcontracting other components of their activities [64].

In host territories, pecuniary externalities of FDI result in the creation of a pool of specialized labor. Depending on the industry, MNEs may create a denser and more diverse labor market. Krugman and Obstfeld [69] argue that a qualified labor benefits both manufacturers and workers. In large and diverse markets, producers are less likely to experience labor shortages while workers are less likely to face unemployment as switching firms and acquisition at convenient market prices gets easier [30].

Blomstrom and Kokko [15] and Dicken [27], argue that one of the most important reasons countries aim to attract FDI is the anticipation to obtain advanced knowledge. The literature identifies and focuses on three potential channels FDI transfers knowledge into the host territory [9].

The first channel of knowledge transfer is through backward linkages. These linkages refer to input purchases of MNEs in host territories. Turok [113] argues that the "quality" of linkages is important to ensure economic development in the host economy. He states that MNEs can establish in a host territory linkages, which can be developmental or dependent in nature. Developmental linkages encourage collaboration, mutual learning, a high level of interaction, and long-term relations with local suppliers. As MNEs strongly interact with local suppliers, they introduce new knowledge, offer technical assistance that improves the quality of local inputs, and provide expertise to promote local R&D activities [112]. In addition, Dicken [27] argues that as linkages become stronger local suppliers find it easier to meet the standards demanded by MNEs and to generate high quality inputs. As more quality inputs become available in host territories, MNEs gradually start to decrease their share of imported goods [60]. On the other side, dependent linkages develop when the purpose of MNEs is to reduce operational costs [110]. MNEs work with local suppliers under short-term contracts. MNEs not only exploit local suppliers but also prevent them to obtain any knowledge or to participate in R&D programs [113]. Furthermore, as in the case of employment FDI indirect contribution to the

GDP of the host territory results from backward linkages subsidiaries establish with local suppliers. Strong ties of MNEs with the host economy allow local suppliers to generate more production, employment, and investments giving rise to multiplier effects in the host territory [120].

The second channel of knowledge transfer is through training of employees. In training programs, employees acquire new knowledge and improve their level of skills [16]. Training programs involve all levels of employees starting from assembly workers to technical professionals and high-level management. The amount and type of training depends on: (i) the industry, (ii) MNE's mode of entry,[4] (iii) size and time horizon of the investment, (iv) local conditions, and (v) responsibilities of the subsidiary [24]. The main purpose of training programs is to transmit knowledge to employees. Employees learn how to produce goods using different methods, how to work in teams, and how to monitor specific operations in the value chain. Employees are exposed to two types of knowledge [109]. The first one is codified knowledge that is expressed in manuals, guidelines, blueprints, software, or hardware. Codified knowledge is transmitted relatively easily across nations. This type of knowledge has enabled multinationals to perform and oversee activities located in different continents [27, 77].

The third channel of knowledge transfer is through demonstration effects. Demonstration effects rely on the argument that local firms including among others suppliers of MNEs imitate/copy technologies introduced by MNEs. Cooperation with MNEs exposes local firms to superior technology appealing them to update production methods [107]. Local firms try to improve production of goods, to use automated machines, to record financial transactions, to access foreign markets, and to generate new business ideas. Demonstration effects occur through spin-offs when local employees previously working for MNEs decide to start local firms based on the knowledge gained while working in MNEs. A well-known example of demonstration effects is the clothing industry in Bangladesh. Employees of locally owned Bangladeshi firms imitated production methods of and ways of running and transforming an enterprise while working with clothing manufacturers in South Korea. Imitation of such demonstration effects enabled the local Bangladeshi firms to transform the clothing industry into a billion dollars industry [36, 65]. Demonstration effects allow for transfer of knowledge that is personalized and possessed by individuals that is virtually impossible to communicate to others through formal mechanisms. It requires direct experience and interaction [105]. Informal exchange of information in social gatherings and group discussions facilitate the transmission of tacit knowledge. Training programs composed of study visits, workshops, and roundtables enable employees to keep up with the latest trends of the sector/industry in which they operate [36, 84].

Another categorization of knowledge transferred in the host territory is between technical and managerial knowledge [102]. Technical knowledge refers to the knowledge gained from local employees that is applied in production and operations of the subsidiary [32, 54]. Managerial knowledge refers mostly to the ability of developing ideas, implementing strategies, interacting with dependent employees in order to

---

[4]Mode of entry refers to whether the FDI is in the form of greenfield or brownfield investment.

succeed in the operations of an enterprise. It covers all aspects of the management of an enterprise ranging from strategic planning and decision making to human, financial as well as logistics and marketing management [2, 43]. A pre-condition for transferring managerial knowledge into the host territory is the appointment of local employees in managerial positions of subsidiaries [37, 42].

Kowalski et al. [66], Banalieva et al. [11] and Young et al. [119] emphasize that a crucial factor for host territories to benefit the most from knowledge transfer is the absorptive capacity. Referring to the degree of host country absorptive capacity is a key factor in the process of knowledge transfer from MNEs to local firms. He argues that knowledge transfer is a complex process that is not automatic, as it requires a significant level of absorptive capacity of local firms. In the process of knowledge transfer, the absorptive capacity in the host territory is a key factor in specifying the level of technicality and complexity of MNE's training programs for their employees [57].

According to the OECD knowledge gap is a key determinant of the host territory absorptive capacity. It refers to the differences in the level and degree of knowledge among different host territories. Bamber et al. [10] indicates that the knowledge gap should not be very wide as local suppliers may not have the capabilities to absorb or copy new technologies transferred by MNEs. In addition, Djordjevic et al. [30], Saini and Singhania [103], Lensink and Morrissey [72], and Moran et al. [80], argue that the process of knowledge transfer has a positive impact only if the host territory possesses a degree of human capital capable of absorbing and using new knowledge and methods. They highlight that the impact of FDI is directly linked to the skills of the labor force. If labor skills are inadequate, host territories cannot assimilate and replicate the knowledge conveyed by MNEs. Martinus and Sigler [78] and Blomstrom and Kokko [15] argue that the level of absorptive capacity determines also the amount and the complexity of knowledge transferred by MNEs. A host territory with a high absorptive capacity attracts more knowledge, which can be relevant to several economic sectors and not only to industries in which MNEs belong [92].

2.5   Evolution of the Subsidiary in the Host Territory

This section is concerned with the evolution a subsidiary experiences during the lifetime it is located in the host territory. By looking into the dynamics of a subsidiary during the "production life" in the host territory, a judgement can be made on the quality of the subsidiary which may go up or down. Upgrading of the subsidiary is an evolution toward a higher quality while downgrading of the subsidiary refers to a reduction in quality [17]. From the host country perspective, the quality of the subsidiary has two components: (i) complexity of activities realized by the subsidiary in the host territory and (ii) embeddedness within the local economy.

2.5.1   Complexity of Activities Realized by the Subsidiary in the Host Territory

Complexity of activities of the subsidiary in the host territory refers to the responsibilities at different levels including the manufacturing process, products, and functions realized within the subsidiary in the host territory [5, 14].

In their paper Enright and Subramanian [38], argue that the complexity in the activities of a subsidiary is determined by the role given from the head office of the MNE. They identify four main subsidiary roles. The first role is that of capability creation. Subsidiaries with such a role undertake extensive research and development activities, set clear strategic objectives, and define functions and responsibilities of the management running the operational activity of the subsidiary [29]. These subsidiaries are regarded as "strategy makers" as their output is used as input in other units of the MNE [3]. The second role is that of capability utilization. Subsidiaries with such a role engage in limited research and development programs. These subsidiaries are called "strategy takers" as their outcome is generated based on inputs derived primarily from other units of MNEs [96]. The third role refers to geographic scope. Subsidiaries take on decision-making activities on a national and regional level. The forth role refers to product-scope [26]. Under this role, the responsibilities and functions of subsidiaries vary considerably. Some subsidiaries are responsible for a number of products while others are in charge only for a single product. The role of the subsidiary in the host territory is assigned based on the characteristics of the group to which the subsidiary belongs, the industry in which it operates, and the economic development of the host territory [84].

2.5.2   Embeddedness of the Subsidiary Within the Host Territory

In the literature, embeddedness is regarded as the integration and attachment of the subsidiary in the host territory through linkages, interaction with local suppliers, created employment, and the level of investments made in the host territory [94]. Phelps et al. [89] refers to embeddedness as working local inputs, capital, public institutions, and human capital in order to create a local network of supplies and a trained working force. Local suppliers include both domestic suppliers and foreign suppliers that have installed production units in the host territory to serve their customers [9, 43]. These linkages create networks of economic interdependence that facilitate the flow of information and knowledge between the subsidiary and suppliers. In addition, Blomstrom and Kokko [15] concluded that a gradual development of backward linkages occurs when (i) more stages are added overtime in the production of goods, (ii) more suppliers enter into a particular industry, and (iii) MNEs prioritize the strategy to attract and create more linkages with local suppliers. A subsidiary is better integrated in the host territory if it creates sustainable employment in the local economy and makes investments that are long term in nature [22, 65, 108].

Referring to Phelps [87], the capacity of institutions in the host territory is of vital importance for subsidiary embeddedness in the local economy. He argues that structured and functional institutions are crucial in designing and implementing effective policies that can maximize the impact of FDI in the host territory [111]. Such institutions can balance the use of financial resources and support services between foreign investors and local firms and can continue to keep the focus also on the existing pool of investors without switching it entirely towards new or potential investors [64].

2.6 Upgrading of the Subsidiary

During its lifetime, the operational activity of a subsidiary changes: it can upgrade or downgrade (see Fig. 2.3). Upgrading occurs when the subsidiary is more embedded within the host economy and/or realizes progressively more complex activities [29]. On the other hand, downgrading occurs when a subsidiary little by little performs fewer and less complex activities and/or weakens the linkages with the host territory [14, 38, 49].

2.6.1 Upgrading in the Sense of Complexity of Activities

Gereffi [44] introduced three kinds of upgrading in the clothing industry: (i) process, (ii) product, and (iii) functional. Reference is made to the clothing industry as a specific objective of this book. He applied it to three levels: firm, industry, and territory. The main focus is upgrading at the subsidiary (enterprise/firm) [24]. According to

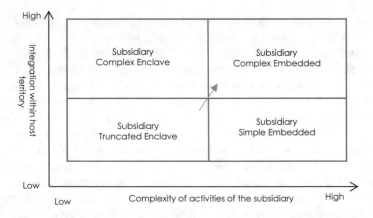

Fig. 2.3 Subsidiary upgrading in the host territory. *Source* Author's drawing

Jones and Wren [60], Kaplinsky and Morris [61] and Humphrey and Schmitz [53], upgrading is achieved by: (i) improving the efficiency of production processes (process upgrading), (ii) adding new product lines that bring improvements in design or technical specifications (product upgrading), and (iii) taking on new functions that demand advanced skills and extensive knowledge (functional upgrading) as presented in Fig. 2.4 [31].

- **Process upgrading** is obtained by applying new knowledge or by rearranging existing systems used in manufacturing of cloths. Innovation has occurred mostly in the pre-assembly stages such as pattern making and fabric cutting [9]. Sewing operations remain labor-intensive as substitutability between labor and capital in the clothing industry is limited. According to Goto et al. [48], in host territories process upgrading of subsidiaries is facilitated from transfers of advanced knowledge obtained from existing linkages with international buyers in production and distribution networks of the clothing industry.

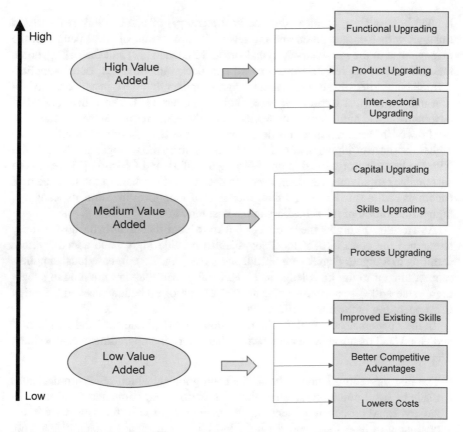

Fig. 2.4 Upgrading of the subsidiary as per value added activities. *Source* Author's drawing

- **Product upgrading** involves a shift into higher sophisticated clothing manufacturing lines that are normally more difficult to produce because of tight technical specifications and expensive input materials used to generate finished goods [84]. For instance, a supplier may experience product upgrading by shifting from manufacturing of casual woven shirts to expensive suits [56].
- **Functional upgrading** refers to shifting towards more knowledge and skill-intensive functions like product design, material sourcing, branding, and marketing [66]. Functional upgrading at the enterprise level depends heavily on the capacity of clothing manufacturers to handle increasingly complex functions and on the willingness of the head office to delegate such functions to its manufacturing subsidiaries located in host territories [20].

2.6.2   Upgrading in the Sense of Embeddedness

Embeddedness (integration within the host territory) of a subsidiary results from linkages with local suppliers and the level of employment of local people in the management of the subsidiary. Moghaddam and Redzuan [79] and Caves [22], argue that embeddedness results from forward and backward linkages. Local suppliers include both domestic suppliers and foreign suppliers that have installed a production unit in the host territory to serve their customers [19]. These linkages when developmental in nature create networks of economic interdependence that facilitate the flow of information and knowledge between the subsidiary and suppliers. The higher the share of inputs obtained from local suppliers the stronger are the linkages with the host territory. In addition, referring to Alfaro and Charlton [2] creation of sustainable employment through appointment of local workers in the management functions of the subsidiary and the generation of long term investments leads to a higher integration of the subsidiary in the host territory.

As a results, the higher the level of global knowledge that flows in the host territory strengthen the capacities of local suppliers to provide high value added services resulting in a higher degree of embeddedness (see Fig. 2.5). In case local suppliers remain limited to the knowledge acquired locally then they are not able to supply high value added services resulting in low degree of embeddedness with foreign firms operating in the host territory [78].

In their papers, Zhou et al. [121], Desbordes and Franssen [26] and Birkinshaw and Hood [14] mention three main factors that determine upgrading of a subsidiary. These factors are:

- The first upgrading factor is head office assignment, which relates to decisions made by head office on the allocation of functions and competences of the subsidiary. Head office assignment may drive the evolution of the subsidiary in the beginning of its activity and when the level of resources and capabilities are not too advanced.

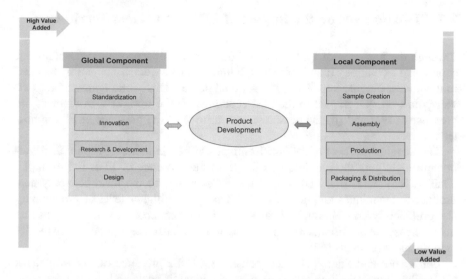

Fig. 2.5 Upgrading of the subsidiary in sense of embeddedness. *Source* Author's drawing

- The second upgrading factor is subsidiary choice that refers to the decisions made by managers on the activities carried out by the subsidiary they supervise. This factor highlights that the subsidiary is part of a network and not just a subordinate to the head office. A subsidiary can move from a position of subordination to one of equality or even leadership. It reflects the fact that many subsidiaries have specialized capabilities on which the rest of the MNE depends. In addition, this factor recognizes that strategic behavior occurs below the management level sometimes in ways that are not actively encouraged by highly ranked management. This autonomous behavior indicates a process of internal growth that is only loosely controlled by the management in the head office. The autonomous behavior appears to be a powerful drive for subsidiary upgrading.
- The third upgrading factor concerns local environment and the influence it has on the head office and/or subsidiary managers regarding competencies of the subsidiary. The argument is that each subsidiary operates under a unique set of conditions to which it has to adapt in order to become effective. Subsidiary upgrading is driven by dynamics occurring in the local business environment as well as by the subsidiary's ability to obtain resources from the MNE. The particular geographical setting and history are responsible for defining a development path that is unique to the subsidiary, which in turn shapes the profile of the capabilities available in the subsidiary.

2.7   Two Schools on the Impact of FDI in the Host Territory

Through the presence of FDI, host territories may attain economic development capital and technology transfer or can fall into an enclave economy getting blocked in the middle income trap. Thus, it has a dual nature. This section presents the two viewpoints on FDI impact in the host economy: (i) the optimists, in favor of the presence of FDI and (ii) the pessimist, not in favor of the presence of FDI in host territories.

The literature in favor of the positive impact of FDI in inducing economic development highlights the multiple benefits from the presence of FDI. For example, Kleibert [65] states that FDI in host territories brings to the local economy new production techniques and new products that cause a higher level of productivity and profitability. In addition, FDI can supply in the host economy better inputs that may increase the performance of local firms and raise customer confidence in locally manufactured products [60].

For Svedin and Stage [112] and Radelet [90], the presence of FDI stimulates the local economy through exposure to international firms using leading-edge technologies, encouraging economic specialization, turning high rates of investment into profitable economic activities, and in providing foreign exchange to finance imports of capital goods, which cannot be produced locally. Use of more advanced knowledge improves the quality of locally manufactured goods and promotes local exports [120]. For the optimists, one of the most important benefits a host territory can obtain through the presence of FDI is knowledge. For them knowledge is an important link between FDI and economic growth the economy specializes in production of complex final goods, in the generation of specialized intermediate inputs, and higher wages [94].

Moreover, in many countries fiscal incentives favor MNEs with a high percentage of domestic inputs, presumably, because they generate more backward linkages. This in turn facilitates creation of a pool of specialized suppliers and a more qualified labor force. According to Elvekrok et al. [37] and Blomstrom and Kokko [15] a host territory exposed to FDI for a long time is characterized by increased linkages as the skill level of local entrepreneurs grows including spinoffs. New suppliers emerge and the local content production goes up, the local efficiency increases by copying operations of the foreign firms increasing survival rates [43]. Moreover, the host territory can benefit from automatic upgrading of the subsidiary when it increases the complexity of activities undertaken in the host territory accompanied with more investments [54].

However, there is an opposing view not in favor of the idea that FDI leads to economic development in the host territory. Bürker et al. [21] suggest that development of the host territory is hindered as foreign firms fail to upgrade. Higher wages in the host territory can be generated through employment of the most qualified workers imposing high wages on foreign and local firms [74]. Another negative impact may arise as foreign enterprises are able to take over in the form of mergers and acquisitions more efficient local firms leaving less productive ones to be owned locally or can acquire a higher share in the local market as locally owned firms stop operating [111].

Coy and Comican [24] and Kaplinsky [62], suggests that foreign firms in the host territory focused on export promotion generate employment mostly of cheap labor creating this way an enclave economy. As such, they do not transfer much knowledge in the host territory and fail to upgrade as undertake only basic functions and processes. As a result levels of national income may depreciate due to global competitive issues, as local firms are unable to upgrade and experience downgrading because they fail to keep up with the knowledge employed by foreign firms [103].

Another negative impact of FDI in the host territory results from the anticompetitive effects inherently from the unequal bargaining power between foreign firms and the government in developing countries [30]. Foreign firms are large, and have a better grasp of the world economy than the government in developing countries that are typically administratively weak and have little information on the functioning of the world economy. Under these circumstances, foreign firms may temporarily exploit location advantages in the host territory without transferring much knowledge generating only a low degree of integration due to the weak linkages with local firms and suppliers [75, 84].

FDI may also prevent creation of a pool of specialized labor in host countries. They offer high benefits and better career opportunities attracting the most qualified workers [88]. Through contractual agreements, FDI may impede the mobility of skilled and trained workers from local firms and institutions limiting the occurrence of demonstration effects in the host territory. Therefore, local firms are left to operate with less qualified labor force constraining their potential for upgrading [26].

However, in the literature there is the view that the presence of FDI in host territories may adversely affect the creation of specialized suppliers. In host territories, FDI tend to be partially financed by host financial markets leading to higher credit costs. Local suppliers with smaller structures than those of MNEs and with limited bargaining power find it difficult to sustain high borrowing costs. They keep postponing necessary investments for further expansion, which at some point in time may cause them to disappear from the market [63].

2.8 Policies to Attract and Foster Upgrading of FDI in the Host Territory

This section presents the main policies followed in several developing countries to attract and upgrade FDI in order to maximize the impact from their presence.

2.8.1   Policies to Attract FDI Within the Host Territory

If developing countries aim to acquire the benefits of the FDI, first they need to attract them. Policies required to attract FDI rely on: (i) creating a stable business environment, (ii) drafting of government incentives, and (iii) advertising the location advantages of the host territory [45].

A host territory appeals to foreign firms if it has a stable environment of doing business. Macroeconomic stability with moderate inflation rates, clear custom procedures on import and export of goods, streamlined construction permits, in availability of uninterrupted power supply, and a stable local currency are indicators of a stable business environment [12, 64]. Another feature is the quality of institutions operating under a well-defined legislative framework that reduces the burden of business operations by having a clear set of procedures (registration, licensing, custom, tax declarations, and getting permits) to be followed by foreign firms. Foreign firms like policies that ensure political stability as they serve as a good indicator of a lower risk of investment [49, 59].

Fiscal incentives include reduction in corporate income tax aiming at encouraging investment in the host territory. Foreign firms are attracted to locations where the fiscal regime imposes minimal costs on cross-border transfer of funds, goods, services, and employees. This includes reduced rates of withholding tax on remittances to their home territory, lower personal income tax or social security reductions for foreign executives and employees [12, 92].

In addition, promotion of location advantages available in host territories has turned into a key policy for attracting foreign firms. In the fast moving world economy, foreign investors may not be aware of all advantages and opportunities in host territories if they are not much advertised [52, 101]. Numerous marketing strategies such as advertising in the media, targeting of potential foreign investors, and holding promotional events can bring more FDI in a host territory. In addition, promotion is also required when state owned enterprises enter into a privatization program, as it will attract more potential foreign investors interested in acquiring the state owned enterprises [49, 59].

The second type of incentive offered to foreign investors is in the form of cost sharing between the government and foreign firms. This type of incentives are in the form of grant or soft loans, sale of land and building to foreign investors at prices below market values, or temporary wage subsidies for employment of local workers [77].

Finally, the availability of a good infrastructure is a key requirement for the presence of FDI in host territories. Investments in roads, ports, airports facilitates the operational activity of foreign enterprises in the host territory. In addition, it facilitates access to neighbouring countries and reduces the delivery time of production designated for exports [78].

2.8.2 Polices to Promote Upgrading of FDI

This section presents various policies that governments in host territories can follow to foster upgrading of inward FDI. The first one is industrial policy and the second is capacity building the skills of the labor force.

Widely advocated by Hausmann et al. [52] and Rodrik [98], industrial policy aims to achieve economic development by stimulating economic activities that result in structural changes and increase productivity in local firms. The main objective of industrial policy is to develop a framework for drafting policies based on the cooperation between the government and firms operating in the local economy [50, 107]. In this relationship, the role of government goes beyond that of macroeconomic stability and extends into economic growth generator undertaking policies drafted based on the dialogue with the private sector including local and foreign enterprises [47]. This way the government can ensure a sustainable development of the private sector in the local economy through manufacturing of goods with more advanced technologies and shifting of resources from old-fashioned economic activities into new economic areas [43, 58].

Recently, industrial policy is oriented towards building up systems, creating networks, developing institutions and aligning strategic priorities of local firms with those of foreign enterprises operating in the host territory [115]. This strategic orientation that aims at the know-how progress of the local economy asks for simultaneous improvements in education, financial and legal institutions, and infrastructure [73, 84].

The new industrial policy is designed with a view of fostering structural transformation patterns by encouraging developmental linkages between local and foreign enterprises creating the potential to accelerate more productive and better jobs [85]. Productive jobs lead to higher levels of income, reduced poverty, improved standards of living, and stronger domestic demand giving rise to higher wages, better working conditions, training and greater social protection [66]. Better jobs, in the sense of having greater developmental impact include those that provide workers with more opportunities to acquire new knowledge and technical competence resulting in enhancing the complexity and diversity of the knowledge base of the labor force, as an essential ingredient for accelerating industrial progress [25].

One of the most common policies to build skills in the labor force is by strengthening the absorptive capacity of local firms in a host territory so that foreign firms can increase linkages with local suppliers allowing them to benefit mostly from knowledge transfer [121]. Improving the absorptive capacity of the host territory in subsidizing local firms to improve their capabilities so they can upgrade their skills and promote cooperation with research and development institutions [116]. Programs spread in the host territory for identification and certification of suppliers are conducted in a transparent manner based on the selection criteria specified by firms who will sign a contract with suppliers that qualify [60]. Another policy applied mostly in small developing countries is to attract small and medium size foreign firms so to gradually assess the capacity of local firms to deal with larger international firms.

Additional policies include public private partnerships that encourage training within enterprises, share of training costs, and research development projects [71]. Finally, public support is in the form of creating industrial parks, reliable infrastructure, and vocational training with curricula designed to encourage cooperation between local and foreign firms operating in the local economy [81].

Quality and diversity of local suppliers available in a host territory appeal to foreign enterprises to produce goods that are more complex and undertake more advanced functions. In order to meet the requirements of foreign enterprises, local suppliers can be trained through various programs undertaken by government institutions that can be financially supported from various incentives [5].

2.8.3 Example of Attracting and Promoting FDI in Various Developing Countries

Several countries have tried to open up cautiously and to attract FDI. Among them are China, Mauritius, Philippines, and Malaysia. Today, East Asian countries that dominate international trade are recognized as the Asian miracle, and are nicknamed as the "Asian Tigers". However, this was not the case some decades ago. These countries undertook a number of structural reforms and strategies to become what they are today. Fernandes and Paunov [39], argue that East Asian countries were able to make a tremendously force like globalization to work for them rather than against them.

Moreover, openness of the economy was not a traditional one, China opened up very gradually. No membership in the World Trade Organization facilitated government intervention. State monopolies of trade were slowly abolished while complex and highly restrictive set of tariffs and licenses to restrict imports were imposed. Chinese government was determined to push economic growth and engaged in export-oriented investments without harming domestic enterprises. Special Economic Zones (SEZ) offering better infrastructure and duty free on imports were able to generate more exports and attract more foreign investments [108]. These investments had a key role in the evolution of manufacturing industries that came to dominate exports. Knowledge was transferred in the country through joint ventures with domestic firms. The government asked foreign firms to have a high content of local components and contents [118]. In addition, foreign enterprise had to work closely with local suppliers in order to ensure among others that technology used in production activities would be similar [29]. Foreign investors that wanted to have a large share in the local market, were not allowed to operate in the country. Local producers benefited from weak enforcement of intellectual protection rights permitting imitation of foreign know-how with little fear of persecution. Moreover, in order to ensure regional development, cities and provinces in China had the freedom to create economic clusters. The government also intervened in the currency market keeping short-term capital flows out while preventing appreciation of the national currency [122]. Finally, China resisted international disciplines and submitted to them only after the economy had a strong industrial base [24, 117].

Even though not geographically located in the East Asia, Mauritius is considered a "miracle" on its own. In creating a strong industrial base, the government founded the Industrial Development Board (IOB) in charge of establishing a policy dialogue with the private sector while promoting new investment opportunities, offering tax holidays to enterprises with the highest prospects of growth, and increasing local employment. The government initiated programs supporting nascent industries [107]. The government stimulated enterprises operating in nascent industries until they were able to compete on their own. Diversification was oriented toward industries meeting domestic needs and those oriented towards exports. Support to export oriented enterprises operating in export processing zones was also included in growth policies. Economic growth in export processing zones was fueled by the presence of foreign investments, exposure to advanced knowledge, investments in capital, and improved entrepreneurship abilities [13].

In the Philippines, the government improved the efficiency of the industrial network by strengthening the cooperative base for shoe production. Among the services offered by the cooperative base were the: (i) the possibility to take out loans at favorable interest rates, (ii) the opportunity to purchase raw materials at discount price, and (iii) the provision of distribution and marketing services to the members of the network [46, 56]. In addition, in Penang, Malaysia, the government supported the creation of industrial clusters in electronics and related industries in order to promote socio-economic development and to attract export oriented FDI [31]. These clusters supported creation of networks between local and foreign enterprises facilitating this way the flow of knowledge and organization of staff trainings. Government initiated programs aiming at stronger industrial capabilities while developing a solid base of exports. In doing so, the government strongly committed to the removal of obstacles preventing private investments like excessive taxation, red tape, bureaucratic corruption, inadequate infrastructure and high inflation. It offered generous subsidies for enterprises to increase investments in modern manufacturing and financial support through loans in the banking sector and tax incentives. In the newly created enterprises, export became a priority since the beginning of their operations [91, 120].

References

1. Aghion P, Howitt PW (2009) The economics of growth. The MIT Press, Cambridge, MA
2. Alfaro L, Charlton A (2013) Growth and the quality of foreign direct investment: is all FDI equal? In: Stiglitz JE, Lin JY (eds) The industrial policy revolution I: the role of government beyond ideology. Palgrave Macmillan, London
3. Altenburg T (2006) Governance patterns in value chains and their development impact. Eur J Dev Res 18(4):498–521
4. Altenburg T, Stamer JM (1999) How to promote clusters: policy experiences from Latin America. World Dev 27(9):1693–1713

5. Ambroziak AA, Hartwell CA (2018) The impact of investments in special economic zones on regional development: the case of Poland. Reg Stud 52(10):1322–1331. https://doi.org/10.1080/00343404.2017.1395005
6. Amin A, Thrift N (2002) Reimagining the urban. Blackwell Publishers Ltd, Cambridge, MA
7. Amoroso S, Moncada-Paternò-Castello P (2018) Inward greenfield FDI and patterns of job polarization. Sustainability 10(4):1–20
8. Asiamah M, Ofori D, Afful J (2019) Analysis of the determinants of foreign direct investment in Ghana. J Asian Bus Econ Stud 26(1):56–75. https://doi.org/10.1108/JABES-08-2018-0057
9. Bahar D, Hausmann R, Hidalgo CA (2014) Neighbors and the evolution of the comparative advantage of nations: evidence of international knowledge diffusion. J Int Econ 92:111–123
10. Bamber P, Fernandez-Stark K, Gereffi G, Guinn A (2014) Connecting local producers in developing countries to regional and global value chains: update. OECD trade policy papers, no. 160. OECD Publishing, Paris. https://doi.org/10.1787/5jzb95f1885l-en
11. Banalieva ER, Santoro MD, Jiang JR (2012) Home region focus and technical efficiency of multinational enterprise. Manag Int Rev 52:493–518. https://doi.org/10.1007/s11575-011-0127-7
12. Bellak C, Leibrecht M, Lieben M (2012) Attracting foreign direct investment: the public policy scope for South East European countries. East J Eur Stud 1(2):37–53
13. Belussi F, Sammarra A (2010) Business networks in cluster and industrial districts. The governance of value chain. Routledge, Oxford, United Kingdom
14. Birkinshaw J, Hood N (1998) Multinational subsidiary evolution: capability and charter change in foreign owned subsidiary companies. Acad Manage Rev 23(4):773–795
15. Blomstrom M, Kokko A (1998) Multinational corporations and spillovers. J Econ Surv 12(2):1–31
16. Blomstrom M (1991) Host country benefits of foreign investment. Working paper no. 3615. National Bureau of Economic Research, Cambridge, MA
17. Blomstrom M, Kokko A (2003) The economics of foreign direct investment incentives. Working paper no. 9489. National Bureau of Economic Research, Cambridge, MA
18. Blomstrom M, Kokko A, Zejan M (1992) Host country competition and technology transfer by multinationals. Working paper no. 4131. National Bureau of Economic Research, Cambridge, MA
19. Borje J, Loof H (2005) FDI inflows to Sweden consequences for innovation and renewal firm location, corporate structure. JIBS/CESIS working paper series, no. 36. Sweden
20. Bruno RL, Campos NF (2013) Reexamining the conditional effect of foreign direct investment. The Institute for the Study of Labor (IZA) discussion paper no. 7458
21. Bürker M, Franco C, Minerva GA (2013) Foreign ownership, firm performance, and the geography of civic capital. Reg Sci Urban Econ 43(6):964–984
22. Caves RE (1996) Multinational enterprise and economic analysis, 2nd edn. Cambridge University Press
23. Colen L, Maertens M, Swinnen J (2008) Foreign direct investment as an engine for economic growth and human development: a review of the arguments and empirical evidence. Working paper no. 16. Leuven Center for Global Governance Studies, Leuven, Belgium
24. Coy AG, Comican HF (2014) Foreign direct investment and economic growth: a time series approach. Glob Econ J 6(1):7–9
25. Criscuolo C, Martin R, Overman HG, Van Reenen J (2012) The causal effects of an industrial policy. Working paper no. 17842. National Bureau of Economic Research, Cambridge, MA
26. Desbordes R, Franssen L (2019) Foreign direct investment and productivity: a cross-country, multisector analysis. Asian Dev Rev 36(1):54–79
27. Dicken P (2011) Global shift: mapping the changing contours of the world economy, 6th edn. The Guilford Press, New York
28. Diez MA (2001) The evaluation of regional innovation and cluster policies: towards a participatory approach. Eur Plan Stud 9(7):907–923
29. Divella M (2017) Cooperation linkages and technological capabilities development across firms. Reg Stud 51(10):1494–1506. https://doi.org/10.1080/00343404.2016.1197388

30. Djordjevic S, Ivanovic Z, Bogdan S (2015) Direct foreign investment and the lack of positive effects on the economy. UTMS J Econ 6(2):197–208
31. Djulius H (2017) Foreign direct investment and technology transfer: knowledge spillover in the manufacturing sector in Indonesia. Glob Bus Rev 18(1):57–70
32. Driffield N, Taylor K (2011) Spillovers from FDI and skill structures of host country firms. Royal Economic Society, London, United Kingdom
33. Dunning JH (1988) The theory of international production. Int Trade J 3(1):21–66
34. Dunning JH (2000) The eclectic paradigm as an envelope for economic and business theories of MNE activity. Int Bus Rev 9:163–190
35. Dunning JH (2001) The eclectic (OLI) paradigm of international production: past, present, future. Int J Econ Bus 8(2):173–190
36. Easterly W (2001) The Elusive Quest for Growth. MIT Press, Cambridge, MA
37. Elvekrok I, Veflen N, Nilsen ER, Gausdal AH (2018) Firm innovation benefits from regional triple-helix networks. Reg Stud 52(9):1214–1224. https://doi.org/10.1080/00343404.2017.1370086
38. Enright MJ, Subramanian V (2007) An organizing framework for MNC subsidiary typologies. Manag Int Rev 47(6):895–924
39. Fernandes AM, Paunov C (2012) Foreign direct investment in services and manufacturing productivity: evidence for Chile. J Dev Econ 97(2):305–321
40. Fleury A (1995) Quality and productivity in the competitive strategies of Brazilian industrial enterprises. World Dev 23(1):73–85
41. Freeman C (1974) The economics of industrial innovation. Penguin Books, Harmondsworth, Middlesex, England
42. Fu X (2008) Foreign direct investment and managerial knowledge spillovers through the diffusion of management practices. SLPTMD working paper series no. 035. Department of International Development, University of Oxford, Oxford, United Kingdom
43. Gachunga MJ (2019) Impact of foreign direct investment on economic growth in Kenya. Int J Inf Res Rev. https://ssrn.com/abstract=3455577
44. Gereffi G (2011) Global sourcing in US apparel industry. J Text Apparel Technol Manag 2(1):1–5
45. Gereffi G, Fernandez-Stark K, Bamber P, Psilos P, DeStefano J (2011) Meeting upgrading challenge: dynamic workforces for diversified economies. Center on Globalization, Governance and Competitiveness, Duke University, Durham, North Carolina, United States
46. Gerson P (1998) Poverty and economic policy in the philippines. Finance Dev 35(3). International Monetary Fund, Washington DC
47. Goto K, Endo T (2014) Upgrading, relocating, informalising? Local strategies in the era of globalization: the Thai garment industry. J Contemp Asia 44(1):1–18
48. Goto K, Natsuda K, Thoburn J (2011) Meeting the challenge of China: the Vietnamese garment industry in the post MFA era. Glob Netw 11(3):355–379
49. Hartwell CA (2018) Bringing the benefits of David to Goliath: special economic zones and institutional improvement. Reg Stud 52(10):1309–1321. https://doi.org/10.1080/00343404.2017.1346371
50. Hausmann R, Rodrik D (2006) Doomed to choose: industrial policy as predicament, blue sky seminar. Center for International Development, Harvard University Press, Cambridge, MA
51. Hausmann R, Rodrik D (2003) Economic development as self-discovery. Journal of Development Economics, Elsevier, 72(2):603–633
52. Hausmann R, Hidalgo CA, Bustos S, Coscia M, Chung S, Jimenez J, Simoes A, Yildirim MA (2014) The atlas of economic complexity: mapping paths to prosperity. https://atlas.media.mit.edu/static/pdf/atlas/AtlasOfEconomicComplexity.pdf
53. Humphrey J, Schmitz H (2002) How does insertion in global value chains affect upgrading in industrial clusters. Reg Stud 36(9):1017–1027
54. Hussain A (2017) Foreign direct investment (FDI) and its impact on the productivity of domestic firms in Pakistan. Pak Bus Rev 18(4):791–812

55. Iamsiraroj S, Ulubasoglu MA (2015) Foreign direct investment and economic growth: a real relationship or wishful thinking? Econ Model 51:200–213
56. Inada M (2013) The effects of foreign direct investment on industrial growth: evidence from a regulation change in China. KIER working papers 856. Kyoto University, Institute of Economic Research
57. Irzova Z, Havranek T (2013) Determinants of horizontal spillovers from FDI: evidence from a large meta analysis. World Dev 42(C):1–15
58. Iyigun M, Rodrik D (2004) On the efficacy of reforms: policy tinkering, institutional change, and entrepreneurship. Working paper no. 10455. National Bureau of Economic Research, Cambridge, MA
59. Jensen O (2003) Investment strategies that really attract FDI. Briefing paper no. 3. Center for Competition Investment and Economic Regulation, United Kingdom
60. Jones J, Wren C (2016) Does service FDI locate differently to manufacturing FDI? A regional analysis for Great Britain. Reg Stud 50(12):1980–1994. https://doi.org/10.1080/00343404.2015.1009434
61. Kaplinsky R, Morris M (2002) A Handbook for value chain. International Development Research Centre, Canada
62. Kaplinsky R (1998) Globalization, industrialization, and sustainable growth: the pursuit of the Nth rent. Discussion paper no. 365. Institute of Development Studies
63. Kaur M, Sharma R (2013) Determinants of foreign direct investment in India: an empirical analysis. Decision 40(1/2):57–67
64. Khalifah NA (2013) Ownership and technical efficiency in Malaysia's automotive industry: a stochastic frontier production function analysis. J Int Trade Econ Dev 22(4):509–535. https://doi.org/10.1080/09638199.2011.571702
65. Kleibert JM (2016) Global production networks, offshore services and the branch-plant syndrome. Reg Stud 50(12):1995–2009. https://doi.org/10.1080/00343404.2015.1034671
66. Kowalski P, Rabaioli D, Vallejo S (2017) International technology transfer measures in an interconnected world: lessons and policy implications. OECD trade policy papers, no. 206. OECD Publishing, Paris. https://doi.org/10.1787/ada51ec0-en
67. Kruger AO (1997) Trade policy and economic development: how we learn. Am Econ Rev 87(1):1–22
68. Krugman PA (1996) Pop internationalism, 2nd edn. MIT Press, Cambridge, MA
69. Krugman P, Obstfeld M (1997) International economics theory and policy, 4th edn. Addison-Wesley, Princeton, New Jersey, United States
70. Kurz HD, Salvadori N (2003) Classical economics and modern theory. Routledge, London
71. Larrain B, Felipe L, Lopez-Calva F, Rodriguez-Clare A (2000) Intel: a case study of foreign direct investment in Central America. Working paper no. 58. Center for International Development at Harvard University, Cambridge, MA
72. Lensink R, Morrissey O (2006) Foreign Direct Investment: flows, volatility and the impact on growth. Rev Int Econ 14(3):478–493
73. Lin J, Chang H (2009) Should industrial policy in developing countries conform to comparative advantage or defy it? A debate between Justin Lin and Ha-Joon Chang. Dev Policy Rev 27(5):483–502
74. Lipsey RE (2002) Home and host country effects of FDI. Working paper no. 9293. National Bureau of Economic Research, Cambridge, MA
75. Lipsey RE, Sjoholm F (2004). Host country effects of inward FDI: why such different answers? Working paper no. 192. Stockholm School of Economics, Stockholm, Sweden
76. Lora EA (2001) Structural reforms in Latin America: what has been reformed and how to measure it. Working paper no. 466. Inter-American Development Bank, Washington, DC
77. Lowe NJ, Wolf-Powers L (2018) Who works in a working region? Inclusive innovation in the new manufacturing economy. Reg Stud 52(6):828–839. https://doi.org/10.1080/00343404.2016.1263386
78. Martinus K, Sigler TJ (2018) Global city clusters: theorizing spatial and non-spatial proximity in inter-urban firm networks. Reg Stud 52(8):1041–1052. https://doi.org/10.1080/00343404.2017.1314457

79. Moghaddam AA, Redzuan M (2012) Globalization and economic growth: a case study in a few developing countries. Res World Econ 3(1):54–62
80. Moran TH, Graham EM, Blomstrom M (2005) Does foreign direct investment promote development?, 1st edn. Institute for International Economics and the Center for Global Development, Washington DC
81. Moran TH (2015) Industrial policy as a tool of development strategy: using FDI to upgrade and diversify the production and export base of host economies in the developing world. In: The E15 initiative strengthening the global. Geneva, Switzerland
82. Narula R (2010) Keeping the eclectic paradigm simple. Multinatl Bus Rev 18(2):35–49
83. Nehru V, Dhareshwar A (1993) A new database on physical capital stock: sources, methodology, and results. Rev Anal Econ 8(1):37–59
84. Newman C, Rand J, Talbot T, Tarp F (2015) Technology transfers, foreign investment and productivity spillovers. Eur Econ Rev 76:168–187
85. Newman C, Rand J, Talbot T, Tarp F (2015b) Technology transfers, foreign investment and productivity spillovers: evidence from Vietnam
86. Newman C, Page J, Rand J, Shimeles A, Söderbom M, Tarp F (2018) Linked in by foreign direct investment: the role of firm-level relationships in knowledge transfers in Africa and Asia. WIDER working paper 2018/161. The United Nations University World Institute for Development Economics Research. https://doi.org/10.35188/UNU-WIDER/2018/603-6
87. Phelps NA (2000) The locally embedded multinational. Area 32(2):169–178
88. Phelps NA (2008) Cluster or capture? Manufacturing foreign direct investment, external economies and agglomeration. Reg Stud 42(4):457–473
89. Phelps NA, Lovering J, Morgan K (1998) Tying the firm to the region or tying the region to the firm? Early observations on the case of LG in South Wales. Eur Urban Reg Stud 5(5):119–137
90. Radelet S (1999) Manufactured exports, export platforms, and economic growth. Harvard Institute for International Development, Harvard University
91. Rasiah R (1994) Flexible production systems and local machine tool subcontracting: electronics components transnational in Malaysia. Camb J Econ 18(3):279–298
92. Reenu J, Sharma AK (2015) Trends and determinants of foreign direct investment in India: a study of the post-liberalization period. South Asian J Manag 22(3):96–98
93. Ritchie BK (2004) Politics and economic reform in Malaysia. Working paper no. 655. The William Davidson Institute, University of Michigan, Michigan, United States
94. Rodrik D (2004) Industrial policy for the twenty first century. United Nations Industrial Development Organization, Vienna, Austria
95. Rodrik D (2011) The globalization paradox, democracy and the future of the world economy. W.W. Norton & Company, New York
96. Rodrik D (2012) We learn nothing from regressing economic growth on policies. Seoul J Econ 25(2):137–151
97. Rodrik D (2013) The past, present, and future of industrial policy. Global Citizen Foundation, Geneva, Switzerland
98. Rodrik D (2008) The new development economics: we shall experiment, but how shall we learn? Faculty research working paper series, no. RWP08-055. Harvard University, Cambridge, MA
99. Romer PM (1990) Endogenous Technological Change. J Polit Econ 98(5):71–92
100. Romer PM (1994) New goods, old theory, and the welfare costs of trade restrictions. J Dev Econ 43(1):5–38
101. Sachs JD, Warner AM (1995) Economic reform and the process of global integration. Brookings Pap Econ Act 1:1–95. Harvard University, Cambridge, MA
102. Sachs JD, Yang X, Zhang D (1999) Trade pattern and economic development when endogenous and exogenous comparative advantages coexist. Working paper CID no. 3. Center for International Development at Harvard University, Cambridge, MA
103. Saini N, Singhania M (2017) Determinants of FDI in developed and developing countries: a quantitative analysis using GMM. J Econ Stud 45(2):348–382
104. Scitovsky T (1954) Two concepts of external economies. J Polit Econ 62(2):143–151

105. Scott AJ (2006) The changing global geography of low technology, labor intensive industry: clothing, footwear, and furniture. World Dev 34(9):1517–1536
106. Scott AJ, Storper M (2007) Regions, globalizations, development. Reg Stud 41(1):579–593
107. Smith N, Thomas E (2017) Regional conditions and innovation in Russia: the impact of foreign direct investment and absorptive capacity. Reg Stud 51(9):1412–1428. https://doi.org/10.1080/00343404.2016.1164307
108. Solomon E (2011) Foreign direct investment, host country factors and economic growth. Ensayos Rev Econ 30(1):4–7
109. Stiglitz JE (1996) Some lessons from the East Asian miracle. World Bank Res Obser 11(2):151–177
110. Stiglitz JE, Uy M (1996) Financial markets, public policy and the East Asian miracle. World Bank Res Obser 11(2):249–276
111. Suyanto, Salim R (2011) Foreign direct investment spillovers and technical efficiency in the Indonesian pharmaceutical sector: firm level evidence. Appl Econ 45(3):383–395. https://doi.org/10.1080/00036846.2011.605554
112. Svedin D, Stage J (2015) Impacts of foreign direct investment on efficiency in Swedish manufacturing. Springer Plus 5:614. https://doi.org/10.1186/s40064-016-2238-x
113. Turok I (1993) Inward investment and local linkages: how deeply embedded is 'Silicon Glen'? Reg Stud 27(5):401–417
114. Vasquez-Barquero A (2002) Endogenous development: networking, innovation, institutions, and cities. Routledge, London
115. Warwick K (2013) Beyond industrial policy: emerging issues and new trends. OECD science, technology and industry papers, no. 2. OECD Publication, Paris, France
116. Warwick K, Nolan A (2014) Evaluation of industrial policy: methodological issues and policy lessons. OECD science, technology and industry policy papers, no. 16. OECD Publishing, Paris, France
117. Yao S (1999) Economic growth, income inequality, and poverty in China under economic reforms. J Dev Stud 35(6):104–130
118. Yin F, Ye M, Yu L (2014) Location determinants of foreign direct investment in services: evidence from Chinese provincial data. Asia research center working paper no. 64. London School of Economics and Political Science, London, United Kingdom
119. Young S, Hood N, Peters E (1994) Multinational enterprises and regional economic development. Reg Stud 28(7):657–677
120. Zanfei A (2012) Effects, not externalities (review). Eur J Dev Res 24(1):8–14
121. Zhou W, Jiang H, Kong Q (2019) Technology transfer under China's foreign investment regime: does the WTO provide a solution. J World Trade 54(3). UNSW law research paper no. 19-59. Forthcoming (2020). https://doi.org/10.2139/ssrn.3436484
122. Zuzuna I, Havránek T (2013) Determinants of horizontal spillovers from FDI: evidence from a large meta-analysis. World Dev 42:1–15

Chapter 3
Global Value Chains and the Participation of Emerging Economies in International Trade

Abstract In the last two decades, structural changes occurring in the global economy have remodeled global production and international trade leading to changes in the organization of industrial networks. This chapter presents an overview of operations and highlights key determinants that affect integration and upgrading of developing countries in global value chains. In addition, this chapter underlines the creation of regional value chains among developing countries as a prerequisite for integration in international trade. The key determinants for participation are measured through time series analysis and based on data sets retrieved from EORA, ICIO and World Integrated Solutions. The results indicate that emerging economies need to attract additional foreign direct investments in order to improve the country's involvement in international trade.

Keywords Global value chains · Regional value chains · Production modalities · Value added

3.1 The Emergence of Production Value Chains

The term "value chain" is used to describe the flow of goods from the first process encountered in the manufacturing of a product through the final sale to the end customer [4]. For De Backer et al. [25], Kaplinsky and Morris [62], a value chain describes the whole set of activities that are needed for a product to go through the inception phase, the production phase, delivery to end customers, and discarding after use. The production value chain includes all activities involved in delivering a product from sourcing raw materials and parts, manufacturing and assembly, warehousing, inventory tracking, order entry and management, distribution across all channels, customer delivery, and management of the information systems necessary to monitor all required activities [10, 70]. A production value chain is characterized by the flow of goods, services, money, and information both within and among suppliers, manufacturers, retailers, and customers. It also includes all types of organizations engaged in transportation, warehousing, information processing, and materials handling. Sourcing, procurement, production scheduling, manufacturing, order processing, inventory management, warehousing, and customer service are all functions

J. Kacani, *A Data-Centric Approach to Breaking the FDI Trap Through Integration in Global Value Chains*, Lecture Notes on Data Engineering and Communications Technologies 50, https://doi.org/10.1007/978-3-030-43189-1_3

performed throughout the value chain [7]. The ultimate goal of the value chain is to meet the demand of the customer more efficiently by providing the right product, in the required conditions, and in the desired time [92, 105]. Each step in the chain generates more value for the customer [85, 99]. Activities in the value chain are not separated as linkages exist between primary activities and supporting activities [83, 97].

3.1.1   The Activity in Global Value Chains

Global production and the generation of services are primarily organized around GVCs. World trade is structured around GVCs making them the most important and the most prevailing organization structure for cross-border transaction among developed and developing countries [96]. As world trade grows continuously, GVCs enter into new dimensions, generating this way new opportunities to reach economic expansion and sustainable growth. GVCs can simply be regarded as the set of activities that employees perform and resources needed in order to bring a product or service from its conception phase until it reaches the end consumer and beyond [46].

The first appearance of GVCs occurred in the end of 1970s and beginning of 1980s when a number of forward looking enterprises transferred production in a number of East Asia countries and Mexico [55]. Such organization in production initially was named "commodity chain". These forward looking enterprises including retailers like JC Penny, Sears, and Kmart together with manufacturing enterprises like IBM, General Motors, Volkswagen had as the ultimate objective to reduce production costs, to shorten processes required for production and to export final goods back to home markets [22, 40, 42, 43]. Such organization led to the concept of GVCs, introduced by Gereffi [42] and it referred to commodity chain of the clothing industry starting with production of raw materials (cotton, wool or synthetic fibres) to final garments used by the end customer. Today, GVCs are used to explain the complete value chain both in manufacturing and in generation of services [49].

In the last two decades, structural changes occurring in the global economy have remodeled global production and trade leading to changes in the organization networks of industries and national economies. The geographical fragmentation of industries and the division of product value in different countries before it reaches the end consumer has resulted in upgrading of processes along GVCs [8]. The increasing fragmentation of production facilitates the connection of geographically dispersed production units into a single industry and serves to identify the ever-changing patterns of trade and production [107]. On a macroeconomic level fragmentation in GVCs helps to understand the interaction and interconnection among national economies. The fragmentation of production has encouraged economic growth in emerging economies leading to an increase in the world demand for goods and services as well as an intensified international trade [13].

One of the main benefits from participation in GVCs is fragmentation of production which lowers costs in international trade. Enterprises face a number of costs starting from the factory where the goods are produced until they reach the final customer [15]. These costs include land transport costs, freight and insurance costs, tariffs and duties, costs associated with non-tariff measures, mark-ups from importers, wholesalers and retailers. Coordination costs is another category present in GVCs as geographically dispersed production units need to be linked and monitored among each other. Coordination and monitoring costs are reduced due to the on-going developments in ICT that have enabled connection among units that are placed at great distance [26].

In addition, according to Kimura and Ando [65] and World Bank [102], transportation and communication costs within GVCs have been significantly reduced as a result of developments in technology including the container and the internet, enabling a smooth and inexpensive coordination throughout the logistics chain. Lower trade costs have been achieved also from implementation of regulatory reforms in transport and infrastructure sectors, playing an important role in the global fragmentation of production.

According to Gereffi and Sturgeon [47], Gereffi and Memedovic [45], a value chain can be a producer-driven value chain or a buyer-driven value chain. In producer-driven value chains, large multinational manufacturers have the main position in coordinating production networks. Producer-driven value chains are typical for capital and technology intensive industries such as automobiles, aircraft, computers, semiconductors and heavy machinery. Profits are generated from scale, volume and technological advances. In producer-driven chains, manufacturers of sophisticated products like aircraft and automobiles are key players not only in terms of their earnings but also in the ability to control suppliers of raw materials, distribution, and retailing. MNEs operating in such chains are most commonly structured as oligopolies [56].

On the other hand, buyer-driven value chains involve large retailers, marketers and branded manufacturers whose main role focuses on setting up decentralized production mostly present in developing countries. Buyer-driven value chains are typical for labor-intensive, consumer-goods industries such as clothing, footwear, toys, and handicrafts. Buyer-driven value chains are characterized by highly competitive and globally decentralized factory systems with low entry barriers in manufacturing [41]. The firms that develop and sell brand named products have considerable control in determining the manufacturing of goods including the volume, the place, production techniques, and the level of profit at each stage [14].

Buyer-driven commodity chains are characterized by highly competitive and globally dispersed production systems. Profits in buyer-driven chains derive from unique combinations of high-value research, design, sales, marketing and financial services that allow retailers, branded marketers, and branded manufacturers to act as strategic agents in synchronizing production outsourced in various countries with customer demand [24, 57].

3.1.2 The Interaction Between Lead Firms and Countries of Allocation of Production

The detailed analysis of GVCs provides insights on the governance between enterprises and other actors that control and coordinate activities in production networks and affect relations between global buyers and global suppliers. Assessment of governance structures in GVCs is necessary to understand the impact of strategies and policies on enterprises and locations in which production occurs. The rationale behind the governance in GVCs is the economic efficiency and competitive advantage based on reducing production costs while promoting open market economies and increased communication among locations involved production. This approach is in line with the advantages advocated by OLI paradigm [19, 103].

Enterprises participating in GVCs continuously alter and redesign their production strategies and location. As such, a strategy that was successful ten years ago might not be successful in the upcoming years. Enterprises in production depend on the disparity that exists in the cost of labor and capital between the countries that are repetitively changing [38]. This brings a key implication that understating operations in GVCs, it is necessary to look beyond industries and transactions occurring in global trade and to focus on reallocation of resources, investments, and the capital taking place between developed and developing countries [16, 79].

The participation of developed and developing countries in GVCs happens through the interaction of lead firms placed in countries that are the main players in the global economy and countries, usually developing (emerging economies), where production is located in order to reap the benefits of lower costs [33].

Lead firms that dominate producer-driven or buyer-driven GVCs evaluate the advantages and disadvantages of offshoring or outsourcing their production in different host territories with respect to cost benefits they will obtain (see Fig. 3.1). Their decisions are mainly based on the overall firm strategy and the products of which they have global competitive advantage [37]. These decisions are periodically reassessed by considering fluctuating determinants such as changes in consumer preferences,

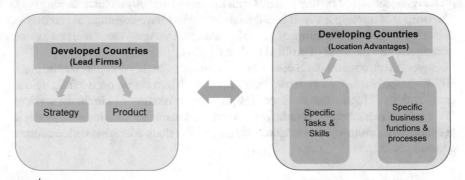

Fig. 3.1 Interaction between lead firms and host territories where production is located. *Source* Author's adaption from World Bank [102] and Johnson [61]

technological improvements, geographical shifts in demand, and risks in host territories. Lead firms within GVCs locate the production cycle in developing countries primarily in order to strengthen their position in global trade, to enter new consumer markets and to reduce the costs of re-exporting goods back to their domestic markets [28]. This strategy quickened the globalization process as lead firms benefited from large pools of low-wage workers, plentiful raw materials and increasing capabilities in manufacturing of complex goods [73].

In developing countries, lead firms look after workers that can perform various tasks while triggering specialization of countries in specific tasks along GVCs. This way lead firms exercise their market power and set entry barriers for enterprises in developing countries where production is outsourced or their subsidiaries exercise their production activities [87]. These barriers arise as lead firms stimulate competition for receiving orders among suppliers located in different host territories to obtain the largest reduction in transaction costs with less focus on increasing the demand for local content requirements or to enhance cooperation with domestic suppliers [63].

A host territory in which the production of GVCs occurs is perceived to obtain specialization in tasks and business functions rather than specific products. This triggers developing countries to compete with each other on the role they will have in GVCs [17]. It also provides insights on the policies developing countries need to undertake in order to eliminate the gap with the entrepreneurial activity compared to key activities occurring in GVCs. Countries can specialize in specific functions such as R&D, procurement, design, operations, marketing, customer services contributing this way to multiple industries and not only to a specific one [36]. When competing with each other developing countries will try to learn specific tasks and skills while mastering business functions and processes. This will enhance their absorptive capacities and will bring developing countries closer to acquire innovation skills aiming at more complex stages of production in GVCs [39].

3.2 Integration in Global Value Chains as an Instrument for Economic Development

Developing countries aim to benefit from the wide spread phenomenon of globalization through participation in GVCs. Participation enables developing countries to enter into regional and global markets through acquisition of foreign direct investments and from amplified participation in the fragmentation of production. This makes it easier for lead firms or MNEs dominating GVCs to locate various stages of productions in different locations and to work with a number of suppliers [34, 35].

Developing countries are the key beneficiaries of the global upsurge in FDI. On average, the inflow of FDI toward developing countries increased on average by 16.4% per year between 2001 and 2016. This average is almost twice the amount of FDI inflows toward developed countries [97]. FDIs is the most common way to link developing countries to GVCs because MNEs are responsible for a large share of trade [98]. Participation in GVCs enables enterprises operating in developing countries to meet standards required to become eligible participants in global markets. They specialize only in specific stages or segments of production without engaging in the whole range of activities embraced by the entire production chain in GVCs [68, 95].

Participation in GVCs allows developing countries to achieve growth in national income through increased trade transactions derived primarily from an increase in the number of exports. From an individual country perspective, participation in GVCs can boost productivity in a number of ways including cross-border division of labor, greater availability of inputs used in production, increased competitions and technology spillovers [52]. In addition, studies undertaken by Ignatenko et al. [58] advocate that developing countries can obtain substantial productivity and economic growth from engaging in export dominated GVCs. Economic growth generated from participation in export oriented GVCs is intensified if developing countries produce intermediate inputs. These intermediate inputs are purchased by foreign investors to be used in the manufacturing of final products. Little impact is obtained if developing countries are focused mainly in assembly of imported inputs and are exploited for the low-cost labor force without affecting the local supply of intermediate goods [61].

Intensive participation in GVCs exposes developing countries to rigorous requirements for highly demanding markets coming from exposure to more advanced knowledge and technology transfer from lead firms to enterprises operating in developing countries [54, 78, 91, 95]. The higher the participation of developing countries through increased imports of intermediate goods and to a lower degree of final imported inputs improves the skills of local enterprises. They get exposed to a higher degree of competition, more complex processes and to the need for compliance with international standards. Local enterprises with the highest performance will be able to enter and enhance participation in GVCs [29].

Active participation in GVCs is beneficial for developing countries for dissemination of knowledge from lead firms to their suppliers operating as local firms in emerging economies. The interaction between lead firms and suppliers puts pressure for upgrading in local firms, pushes them to comply with international standards and creates a sound innovation system that reduces the complexity and facilitates occurrence of transactions with lead firms Pietrobelli and Rabellotti [89].

Developing countries can generate the highest benefits from participation in GVCs by having a set of pre-existing capabilities that are critical not only to influence the initial decision of lead firms to locate production but also for upgrading in GVCs. This condition refers to the absorptive capacity of the host developing country. The domestic absorptive capacity is particularly linked to the quality of labor, institutions, and infrastructure [48]. Among other things, the presence of highly skilled labor permits a positive interaction between imported capital goods and firm productivity [6, 56, 109]. Therefore, having an adequate pool of skilled labor can facilitate the absorption of new technologies and create a climate for further development and sustainable economic growth. The technical capabilities of local enterprises in developing countries also determine their position in GVCs with those having the highest capabilities being able to upgrade while those with the lowest technical skills remain within the segments that require only low skilled labor and including assembly in clothing manufacturing [20].

3.2.1 What Drives Integration in Global Value Chains

Developing countries seek not only to enter into GVCs but also to move upwards along GVCs for long term expansion. Integration into GVCs goes into three main phases (see Fig. 3.2). The first phase is entrance into GVCs by attracting of FDI and establishment of forward linkages rather than backward linkages [21]. The competitive advantages of developing countries at this stage are mainly resources in terms of low-skilled and low-paid workers or natural resources. The second phase is expansion within the operational activities in GVCs. At this phase, developing countries are able to perform processes that are more complex and to diversify the provision of services and skills to lead firms that dominate GVCs [9]. In addition, involvement in more complex and value added functions results from an increased absorptive capacity of local firms operating in developing countries that interact with lead firms in GVCs. The third and final phase of integration in GVCs brings sustainable growth for countries that started as low-income and have achieved a steady growth that goes beyond technological upgrading in production and focuses more on promoting social upgrading and cohesion together with promotion of environmental sustainability and green economy [104, 102].

It is important to note that not all developing countries manage to enter GVCs and out of those that enter only a few make it up to the third phase. As such it is of interest to look into the determinants that influence integration and those that affect upgrading in GVCs.

To start with, trade liberalization speeds up integration into GVCs as trade lies at the heart of operations in GVCs. Trade liberalization facilitates the international movement of goods among geographically dispersed locations as tariff and non-tariff barriers are relieved through multilateral and bilateral agreements [77]. In an

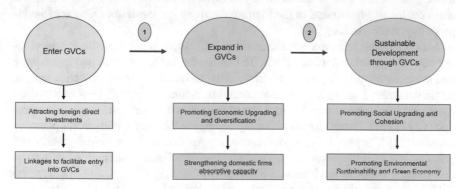

Fig. 3.2 Phases of integration in GVCs. *Source* Author's adaption from World Bank [104]

already largely liberalized trade environment, there is available for developing countries not only free movement of goods but also a simplification of knowledge transfer through the import of technologically advanced capital goods used to perform complex processes in manufacturing [76].

Vicinity to an economically developed center or country is another determinant that accelerates integration in GVCs. Vicinity translates into lower transportation costs and fast delivery while sharing a common culture, social norms and institutions reduce transaction costs. Similar success stories include Republic of Korea and Taiwan that due to geographical were included in the GVC lead by Japan, becoming this way global leaders on their own [74]. Vicinity can be approached also in the creation of special industrial zones with the ultimate objective to integrate into GVCs by offering a more favorable regime than the rest of the country especially for foreign investors that can also enjoy the benefits associated with clustering. Special benefits in low wage countries, predominantly in East and Southeast Asia and in Central America, offer generous investment incentives, including but not restricted to lower income tax, lower or no import duties, subsidized utilities, and cost of land in special zones [65]. Location within a specific geographical zone may increase access to improved infrastructure, more specialized labor and other shared resources such as research institutes that encourage formal and informal knowledge exchanges among participants. In China and more recently Viet Nam have, with some success, created special zones to attract FDIs Lanz and Piermartini [71]. However, even though special economic zones may be effective in attracting foreign enterprises, particular attention needs to be given to participation of domestic firms in order to enhance linkages with FDIs as to advance participation in GVCs.

Building up, the manufacturing competitiveness is the first requirement for developing countries to participate in GVCs and afterwards to move into more complex processes or industries. Integration in GVCs strengthens industrial competitiveness through technology transfers and other types of learning, but also the reverse is true as developing an initial competitive manufacturing industry may be a prerequisite for successful GVC integration [64]. The value-added share in manufacturing is a useful performance indicator when assessing structural change in low- middle-income countries. This is based on the assumption that manufacturing acts as the main engine of growth due to the high productivity growth that can be achieved if proper upgrading occurs [30].

Manufacturing has three characteristics that make it a prerequisite for developing countries to integrate into GVCs. To begin with, the knowledge and technology embodied in manufacturing is relatively easy to transfer across countries and, in particular, from high to low income countries. Manufacturing is characterized by economic convergence over the long run as countries that start with lower labor productivity in manufacturing tend to display higher productivity growth over upcoming decades [53]. The productivity convergence happens notwithstanding of government policies, role of institutions, or geographical location. Better functioning institutions and friendly business environment policies can facilitate the speed of convergence [91]. The second characteristic is that manufacturing is tradable and can be easily exchanged. There is always an external market for provision of manufacturing services, and manufacturing in a developing country does not need to be constrained by domestic demand and income. As such productivity in manufacturing can go up even if productivity in the rest of the economic activity remains low and does not change much [81]. Thirdly, manufacturing is based on low skilled labor that is found in abundance in developing countries. Manufacturing sites of simple garments, toys, or even automotive vehicles could operate by employing low-skilled workers with basic manual agility. Because it is labor intensive manufacturing is especially suitable for developing countries to integrate into GVCs without running into adversities encountered with more intricate operations part of GVCs [80, 81]. These three characteristic make manufacturing an easy to employ instrument to obtain high income in developing countries. Manufacturing remains an exchange mechanism in international trade and within GVCs, however comparative advantage of low-income countries in standard manufacturing tends to dissipate as skills and operations become more technologically advanced. Therefore, skill biased technological change requires developing countries to direct substantial investments in human capital and acquisition of new technology if they want to move forward [88].

The absence of competitiveness of domestic firms poses a problem for integration into GVCs as basic industrial capacities are a prerequisite to establish initial linkages in global trade. For example, in Viet Nam, locally owned enterprises have low participation rates in GVCs because foreign owned enterprises operating in Viet Nam are not able to identify local suppliers to match their quality of international standards [86, 93]. This situation results in much stronger production linkages between foreign enterprises in Viet Nam are much stronger than between foreign and locally owned enterprises.

Integration into GVCs requires creation of manufacturing sites equipped with large physical investments (buildings, equipment, machineries, etc.) and operated by specialized human capital (more educated employees, training programs, study visits, etc.) Easy access to financial capital stimulates and increases the possibilities for developing countries to participate in GVCs [66]. Countries that have consolidated stronger financial markets including better access to credit are more attractive for MNEs to locate a segment of production chain where they can easily perform financial transactions [11]. A study from Manova and Yu [75] concludes based on Chinese customs and firm-level data, that credit constraints restrict firms to remain into low value-added stages in GVCs such as pure assembly as more complex processes in GVCs involve higher setting up costs and require more working capital.

Institutions are particularly relevant for integration into GVCs as they influence transactions and the operational activity required for a good performance in GVCs. Institutions that facilitate contract enforcement and mitigate contractual frictions are characteristics of countries that tend to export more in GVC intensive manufacturing industries [15, 71]. Property rights, are used as a measure for the quality of legal institutions and a major determinant for forward integration into extra-regional GVCs. Transaction costs are affected not only by formal institutions, the rule of law and contract enforcement, but also from informal institutions. Specifically, informal institutions such as reputation and trust achieved through repeated interactions, can replace for formal institutions and facilitate trade in GVCs [84].

Another factor that significantly affects integration in GVCs is the level of skills that explains to a certain extend how enterprises in developing countries can turn from regional to global suppliers. This is achieved as skills positively influence creation of forward linkages [108]. The level of skills directly influences industrial competitiveness that is consistent with overall principles of operations in GVCs where backward linkages are related to low skilled tasks such as assembly, while forward linkages are highest for advanced countries such as the United States, Japan or Finland. As the level of skills in the workforce of a developing country improves, more forward linkages are generated enabling the country to move up in GVCs by contributing with domestic value added in provision of services together with value added in manufacturing of exports [1].

Kummritz et al. [69] explore policy factors that affect integration and upgrading in GVCs as measured by domestic value added. They conclude that factors such as connectivity, education skills, and standards have a stronger association with domestic value added through forward linkages than through backward linkages. Countries that previously were at the entry phase of GVCs with the acquired knowledge and skills have turned into facilitators of increased competitiveness in the region they operate [2]. More recently, Thai firms have been improving the skills and increasing the competitive advantage of Cambodian, Lao and Myanmar workers by establishing production facilities in border economic zones that are part of regional value chains [63].

Time is another factor that affects fragmentation and location of production units in GVCs as it is directly correlated with trade costs and the success of operations in GVCs [28]. Time related trade costs have two components: the speed that it takes to deliver and the ability to deliver on time. Fast delivery is of high importance for GVCs that are characterized by demand fluctuations and quick technological change. Ability to deliver is more critical for GVCs that are characterized by high inventory cost and the delivery of intermediate products used as inputs in production of final goods. Schaur estimates that each day in transit is equivalent to an added tariff of between 0.6 and 2.1%, with intermediated goods being affected 60% more than other goods.

Another factor that is of high importance in terms of GVC participation is the type of ownership of the firms. For example, in China, integration in GVCs is correlated with ownership of the enterprise. More than 90% of fully export-oriented enterprises are foreign-owned or foreign-affiliated. On the contrary, only 71% of Chinese owned enterprises that are export oriented are directly or indirectly integrated in GVCs [67].

More generally, there is no single determinant for integration and active participation in GVCs, but rather a number of determinants must be jointly considered. Developing countries need to upgrade their educational systems and technical training, improve their business environment, and enhance their logistics and transport networks in order to gain full benefits from GVCs [51].

3.2.2   Upgrading in Global Value Chains

Developing countries are in constant competition for foreign investments and con-tracts with global brand owners, leaving many suppliers with little advantage in the chain. The result is an unequal division of the total value-added activities along GVCs in favor of lead firms (see Fig. 3.3). Gereffi and Fernandez-Stark [46], Gereffi and Sturgeon [47] and World Bank [104] have identified the following value adding activities in the clothing industry also referred to as the smiley curve of GVCs. The level of value added activities corresponds also to the three phases of integration in GVCs.

- **Innovation**—It refers to activities related to introducing new technologies that impact the production cycle in GVCs and affect the geographically distributed units that participate in GVCs.
- **Brand**—It refers to creation and possession of products or services that dominate activities in GVCs.
- **R&D**—It refers to activities related to improving products restructuring production processes, identifying new markets, and complying with customer demand.
- **Design**—It refers to activities used to attract attention, improve product perfor-mance, cut production costs, and may increase the competitive advantages of the product in the market.
- **Purchasing/sourcing**—It refers to inbound processes involved in purchasing and transporting textile products. It includes physical transporting of products, as well as managing or providing technology and equipment for supply chain coordination. Logistics can involve domestic or overseas coordination required for goods to reach final consumers and for intermediate goods to reach assemblers.

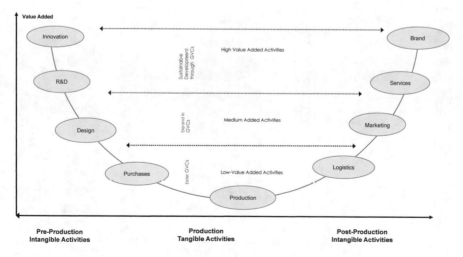

Fig. 3.3 Value added activities according to the phases of integration in GVCs. *Source* Author's adaption from Gereffi and Fernandez-Stark [46], Gereffi and Sturgeon [47] and World Bank [104]

- **Production/assembly/(cut, make, trim-CMT)**—It refers to processes like cut, sew, woven, or knit fabric and yarn. Manufacturers usually acquire raw materials like knitted fabric or yarn from specialized suppliers as agreed with lead firms. After goods are manufactured, they are distributed and sold via a network of wholesalers, agents, logistics firms, and other companies responsible for value-adding activities outside production.
- **Marketing and sales**—It refers to activities like pricing, selling, and distributing a product, including also branding or advertising.

Integration of developing countries in GVCs is a pre-condition but not a guarantee for upgrading and scaling up into higher value added processes and tasks within the production cycle of GVCs. Upgrading in GVCs means moving from production of commodities into more complex manufacturing until developing countries and enterprises are able to provide full service within GVCs (see Fig. 3.4). These steps are complementary with the three phases required for integration into GVCs, more specifically (see Fig. 3.4): (i) production of commodities occurs at the entering into GVCs phase, (ii) manufacturing corresponds to expansion into GVCs phase, and (iii) provision of services happens when sustainable development is achieved within GVCs [69].

To begin with, involvement in GVCs starts with value added processes corresponding to production of basic commodities that require engagement of a lowskilled labor force with the main tasks being assembly, packaging and distribution [95, 99]. The next step is to get involved into processes and activities that are regarded as medium value added and refer to manufacturing that requires more advanced skills than commodities. IT includes tasks like design, management of the logistic chain, marketing and provision of customer services. The last step in upgrading is provision

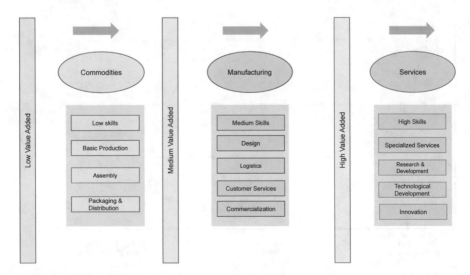

Fig. 3.4 Steps for upgrading in GVCs. *Source* Author's adaption based on World Bank [104] and Fernandes et al. [34, 35]

of services in GVCs resulting in a close cooperation with lead firms. Provision of highly technical services including research and development, enables participation in innovation centers jointly with lead firms to work on innovative products and on potential global technological development [87].

Integration into GVCs does not guarantee that upgrading will occur. Development of linkages (particularly forward linkages), with lead firms in GVCs, can be a significant source of knowledge, leading to successful upgrading. While development linkages within GVCs are a major prerequisite, the benefits are not automatic. Countries most likely to upgrade are those benefiting from participation in high quality export segment in GVCs [75, 81]. The benefits from integration in GVCs are highly volatile and mirror different trajectories showing that only few countries have made considerable progress over time in upgrading and becoming global or regional suppliers of services [38, 48].

To continue, despite the improvement in technical skills and investing in R&D are important drivers of upgrading within GVCs, the outcome varies depending on the phase of integration and the nature of involvement in GVCs [19, 53]. The level of skills is significant when it comes to creation of forward linkages resulting in the production of more advanced products while it is not a requirement for creation of backward linkages such as provision of assembly services. A developing country can expand its overall value added in GVCs while, still remaining in manufacturing and switching basic assembly work with other tasks like design, purchases, marketing, etc. [4, 8]. Another strategy is to extract more value added from production through improvements in productivity, relying on skill formation and innovation through continuous investments in developing technological capabilities [1].

When considering upgrading, participants in GVCs need to carefully look into various GVCs that have different growth potential [9]. Countries and enterprises participating in GVCs located in Asia differ significantly in terms of size, organization, degree of internationalization and rate of growth from enterprises participating in GVCs located in South America or Africa. In addition, different industries experience technological change differently creating distinctive opportunities for upgrading. In GVCs characterized by fast technology cycles like mobile phones or automobile, it may be more difficult to upgrade and catch-up than in chains that rely on slower growing technologies like the clothing and the toy industry [23, 25].

3.2.3   Risks for Developing Countries from Participation in Global Value Chains

In spite of the benefits that developing countries aim to obtain from participating in GVCs, there are also risks associated from involvement in GVCs. When lead firms engage in fragmentation of production mostly for exploitation of unskilled, low-cost labor or natural resources developing countries attain only low levels of upgrading and backward linkages [17, 81]. To continue, even though compliance with international standards stimulates competitiveness, standards can serve as barriers for local enterprises to supply intermediate or final goods because lead firms can decide to reduce the local component in production and secure most of intermediate inputs from different sources [24, 31].

Recently, it looks like that exports are generating less employment than before in developing countries and participation in GVCs is not facilitating. This trend challenges the idea of trade as a driver of economic growth in developing countries. For example, in Ethiopia and the Philippines, the percentage of export-related jobs is halved while the decline in Thailand is visible but not still not at the highest concerns [30, 41]. This indicates that technological transfers and structural benefits associated with exports are not distributed to developing countries. Another concern relates to the wage levels in developing countries as higher level of exports tend to be associated with higher wage levels. This indicates that the gains related to a high level of exports in international trade remain within a limited number of firms that closely work with lead firms within GVCs [49, 68].

To continue, even though acquisition of new technologies through participation in GVCs is of major importance as new technologies can adversely impact economic growth in developing countries. New technologies make it difficult for developing countries to use the labor cost advantage to offset the technological disadvantage, cutting down their ability to substitute low skilled labor for more complex processes in the production cycle of GVCs. The inability to align with technological progress results in the loss of low-skilled jobs in manufacturing and in the increase in the wage differential between skilled and low-skilled workers [32, 74]. Consequently, a number of developing countries may find themselves stuck in a low-wage, low-productivity trap with a high possibility of losing long-run comparative advantage associated with lower labor costs [61].

Finally, participation in GVCs has resulted in uneven distribution between countries and among enterprises participating in GVCs. Larger countries tend to be more favored in GVCs as they usually are characterized by modern infrastructure, have enterprises with higher chances of scaling up in GVCs and a growing domestic demand for goods and services [77, 90]. This trend can disrupt globalization and calls out for policy options that can improve outcomes for developing countries disadvantaged from size in terms of employment and income and to identify appropriate instruments that can reduce risks associated with uneven gains in GVCs [93].

3.3 Regional Value Chains as a Road for Integration in Global Value Chains

Developing countries can speed up integration into GVCs through the establishment of regional value chains based on an unified policy approach for domestic integration [96, 97]. Efficient participation in GVCs is built on well established and connected domestic markets. Regional value chains are effective channels between domestic markets and GVCs with the ultimate objective to expand markets and to increase the operational scale. Regional markets serve as instruments for developing political, social and strategic relations among the countries that are part of a regional value chain [15, 108].

Successful regional value chains are created based on careful consideration of demand and supply side strategies, preferential trade agreements, economic cooperation agreements, and regional production networks [5, 26]. The demand-side logic of regional integration emphasizes the importance of market size, market access, attraction of FDI and economies of scale as key elements for serving on larger markets. On the other hand, strategies on the supply side target provision of improved production and processing services in order to be able to supply higher value added exports from the region. Costa Rica, is a successful example of effective supply side strategies on improving production capacity and skills by partnering with Mexico on training programs to improve the skills of the working force [76]. In addition, Nicaragua, whose clothing manufacturing enterprises have traditionally purchased textiles from East Asia, is consciously pursuing supply arrangements with textile firms in Honduras and Guatemala. Other examples of successful regional value chains include East Asian supply base that is created for China's inputs in electronics used in exports of smart phones (United Nations Industrial Development Organization [44, 99] and sourcing of mineral inputs in Sub-Saharan Africa [50, 63], indicating a rise in the importance of regional value chains for integration in GVCs.

A number of factors can explain the emergence of regional value chains. To start with, geography helps regionally clustered host territories to stimulate the interaction between countries and enable quick delivery times and provision of quick solutions to order problems that can be solved within a day [21, 103]. In addition, trade liberalization programs at a regional level have reduced trade barriers among countries within geographical proximity. Information costs provide another explanation for the significance of regional value chains. Information costs that go up with distance, limit the ability of firms to create a regional network of producers and exporters [25, 81].

As expected, developing countries that have a high fraction of trade exchanges supported by trade agreements are better integrated into a regional value chain with both backward and forward linkages [4, 11]. Regional trade agreements are relevant for creation of regional value chains as they can facilitate for members integration in GVCs while can adversely affect developing countries that do not belong to any regional trade agreement.

Regional value chains are strongest in the European Union single market and in Southeast and East Asia [72, 102]. The importance of regional value chains has been steadily increasing. For example, in 2011 intra-regional backward linkages accounted for nearly 12% of Asian manufactured exports, while backward linkages in GVCs were about 17%. However, forward linkages are more dominant in GVCs rather in regional value chains [91].

3.3.1 Western Balkans as a Regional Value Chain

During the last 25 years, exports have significantly contributed to the economic growth in many small emerging economies. However, the Western Balkans[1] (WB) region, consisting of small open market economies, has not taken full advantage of this driver of growth and convergence. Countries in the WB are not well integrated into the dynamic European GVCs [104]. Moreover, trade among WB countries is inadequate and mostly based on bilateral agreements rather than organized in a cluster.

Although trade exchanges in the WB region have gone up over time, this is attributed mostly to the exports of labor intensive and low value added products. The long transition process and the slow progress in key structural reforms have restrained the competitiveness of the WB region despite of low wages compared to European member states [100]. In order to strengthen its competitiveness the economic-growth model needs to be oriented toward an increase in exports and better integration in GVCs with no need to build expertise in all industries or goods.

Backward linkages are more common between WB countries and GVCs, as these countries serve mostly as assembly centers for low value manufacturing products. WB countries have fewer trade links with Germany the most GVC hub in Europe while the majority of them strongly trade with Italy a country trapped for years in the euro area crisis inhibiting the export growth required for integration into GVCs [60, 100].

The potential of WB countries to become an united economic area in order to transfer significant capital and trade in the region is not well utilized mainly due to how regional trade agreements are implemented. WB countries are members of the Central European Free Trade Agreement (CEFTA)[2] on goods trade and investments, under which members enjoy zero tariff on trade. However, the low levels of integration in regional trade among WB countries come from bilateral tariffs, lack of recognition of production standards and not unified customs procedures [34, 35]. The WB countries have thinner trade agreements with 38 countries (mostly European) with few legally enforceable provisions. The WB region can obtain significant gains from targeted reforms especially in trade. To attract investment from European

---

[1] The Western Balkan Region is composed of Albania, Bosnia Herzegovina, Kosovo, Montenegro, Republic of North Macedonia.

[2] The Central European Free Trade Agreement (CEFTA) is a trade agreement between non-EU countries, mainly located in Southeast Europe. It was founded by Hungary, Poland and former Czechoslovakia in 1992.

It evolved to CEFTA 2006 with six new parties from Southeast Europe. The necessary ratification processes of this new CEFTA were ratified between parties during July and November 2007. The speed of the process showed the importance that CEFTA had regarding its members. The parties of CEFTA 2006 are Albania, Croatia, Macedonia, Moldova, Montenegro, Bosnia and Herzegovina, Serbia and the United Nations Interim Administration Mission in Kosovo (UNMIK).

As foreseen in the agreement, Romania and Bulgaria left CEFTA after their entrance in the EU in 2007. Followed by Croatia in 2013.

The leadership of CEFTA changes on an annual basis. It is responsible for calling, chairing, and reporting on all meeting of the Committee.

economies the WB countries can raise the level of institutional readiness and labor skills that increases trade transactions in GVCs by in 6–12% points [59].

According to a recent IMF Survey [59] undertaken in WB countries the respondent enterprises indicate that constraints such as skills shortage and low institutional quality, affect the level of exports and participation GVCs. Enterprises participating in GVCs have developed backward linkages with the EU and OECD countries while their forward linkages are limited, as they mainly focus in the assembly of final products. Enterprises participating in this survey are attracted to the WB region due to geographic proximity to investors, favorable taxation, cheap labor, low tariffs. However, they have identified a few obstacles such as political instability, poor trade logistics, shortage of skills and inefficient institutions [60, 100].

Consumption-based growth model characterizes the structure of imports in WB countries. Consumption of imported goods has a large share ranging from 33 to 45% in Serbia and Albania respectively. The manufacturing sector consists of labor-intensive products, apparel, unprocessed foods, leather and it prevails mostly in Bosnia and Herzegovina and Albania [100].

In order for WB countries to utilize and absorb advanced technology in imported capital goods an upgrading of skills through better education and labor market policies needs to take place. A reconsideration of policies and structural reforms is needed to help with the slowdown of income convergence between the Western Balkan countries and the EU in the forthcoming decade. Despite its favorable position in the largest trading bloc in the world, exports in the WB region have played a subdued role in its economic growth [59].

In order to integrate into GVCs and experience its benefits, an improvement in infrastructure and in the labor skills is required in WB countries, accompanied with deepening of trade agreements. Based on the findings of this paper, increasing links in GVCs will raise the GDP level of WB countries by 3–10%. Foreign investment, regulatory reforms, improvement in institutions are required for integration into GVCs. Deep trade reforms, improvements in infrastructure and education are also required to enter Europe's production and services networks. Infrastructure of WB needs to be seriously upgraded in order for this to happen [5].

3.3.2 Albania and Its Integration in GVCs

Albania is a small country located in the Western Balkans Peninsula in South East Europe and one of the oldest nations of the region. It is also known as the "Land of Eagles". It has a surface area of 28,748 km^2 and a population of 2,898,293 million inhabitants as of the latest census of National Statistical Institute of Albania (INSTAT)[3] in 2011. The Albanian population has a homogenous ethnic composition with the presence of Greek, Romanian, Bulgarian, and Macedonian minority groups.

---

[3]The information was obtained from www.instat.gov.al.

It shares borders with Montenegro, Kosovo, FYR Macedonia, and Greece. In the literature, Albania is considered as a "gateway" between East and West [27]. After the 2000s, Albania intensified its efforts towards European Integration, one of the main objectives of the country.

In 2006, Albania signed with the European Commission the Stabilization and Association Agreement starting the road toward membership in the European Union. In 2012, the European Commission recommended the EU candidate status and it is currently working for opening of negotiations for EU accession. In 2009, Albania gained full membership in the North Atlantic Treaty Organization (NATO), getting ahead of neighboring countries like Serbia, Montenegro, Bosnia Herzegovina, and Republic of North Macedonia.

After the Second World War, Albania joined the East Communist Block, which was followed by 45 years of dictatorship and an extremely self-secluded policy. During the dictatorship regime, the Albanian economy was ruled by the concept of state ownership in all means of production, including agricultural land and by censuring all forms of private property or entrepreneurial activities. This regime left the country with no room for entrepreneurship and with very limited knowledge on how to operate in the world economy. The fall of dictatorship in the early 1990s has marked the re-birth of the country together with the creation of a democratic society, establishment of the market economy, and promotion of an open economy. After more than 45 years of seclusion and communist regime, Albania inherited an outdated and almost inexistent heavy industry (metallurgy, chemicals, and petroleum) on which considerable investments are made during the dictatorship. Many coal, chrome, and other mines closed or operated on very limited capacity while the oil and gas industry ceased to function. In addition, small fabrics and large industrial plants including the processing of agriculture crops were not functional anymore [18, 82].

Since the early 1990s Albania has experienced an important economic transformation, which has significantly reduced poverty and has placed Albania into the ranks of middle-income countries (see Table 3.1). Within the last three years domestic demand increased despite political tensions. Private consumption increased due to improved employment, higher wages and growth in consumer credit. Better credit conditions and government infrastructure spending pushed up investment. In the country, job creation strengthened and unemployment declined despite lower growth. Growth in jobs increased due to new jobs created in agriculture and the service industry. In addition, real wages increased, mainly in the service industry.

Inflation has declined since the Bank of Albania has held its policy rate at a record low of 1% since June 2018. The banking sector is in very good state, being profitable and well-capitalized. Loan portfolios improves as the Bank of Albania continued to restructure non-performing loans (NPLs) of large borrowers. Increased economic dynamism can gradually close the output gap causing an accelerated growth of 3.5%. Although exports have experienced positive trends in Albania, they will moderate due to a stagnant growth in global economy. Growth will also be affected by investments, fueled by private investments and public projects, only if reforms still continue to progress [101].

Table 3.1 General indicators on the economic development of Albania

Indicator	Value	
Surface	28,748 km^2	
Inhabitants in 2017	2,876,591	
GDP in 2017 (Mln/EUR)	11,546	
GDP per capita in 2017 (EUR)	4024	
Economic growth (average of yearly GDP growth 2013–2017) in (%)	2.42%	
Composition of GDP 2017	in Mln/EUR	in %
Agriculture, forestry, and fishery	2195	19.01%
Extracting industry	277	2.40%
Manufacturing industry	709	6.14%
Construction	1059	9.17%
Trade, transport, hotels, and restaurants	2090	18.10%
Information and communication	356	3.08%
Financial and insurance activities	289	2.50%
Real estate activities	650	5.63%
Professional, technical and scientific activities	339	2.94%
Administrative and supporting activities	386	3.34%
Public administration, security, obligatory social security	520	4.50%
Education	492	4.26%
Health and social work	341	2.95%
Arts and entertainment	125	1.08%
Other household and service activities	184	1.59%
Transport and inventory	368	3.19%
Accommodation and alimentary services	251	2.17%
Unemployment rate (%) in 2017	7.50%	
Inflation rate (%) in 2017	2.00%	
Exports (% of GDP) in 2017	17.75%	
Imports (% of GDP) in 2017	40.14%	
FDI stock (Mln/EUR)	2013	2017
	2122	6455
FDI stock in the manufacturing (Mln/EUR)	2013	2017
	−185	591
Clothing industry (% GDP) in 2017	2.10%	
Clothing industry (total employment) 2016	24,700	
Public debt (% of GDP) in 2017	70%	

Source INSTAT, Bank of Albania, Ministry of Finance

In mid-1990s, the Government of Albania (GoA) launched a new economic agenda to stabilize the macroeconomic environment and to sustain a major transformation of the economy during which economic resources moved from agriculture to construction and services. The expansion of these sectors was driven by a strong domestic demand for modern apartments and the absence of alternative investments. In the future, as the demand for residential constructions decelerates, it is crucial for the economic performance to shift investments from a risky sector like construction into more promising areas for sustained economic growth such as manufacturing. One way Albania can revitalize the manufacturing industry is by attracting foreign direct investments (FDI) and benefit from their presence in order to move in the ladder of economic development (see Fig. 1.2). According to the United Nations report on FDI in Albania (2011),[4] in labor-intensive industries such as clothing and footwear the country has attracted a large number of investments coming mainly from Italy, Greece, and Germany. Moreover, this report states that Albania had a 30% increase in the inflow of FDI mainly in manufacturing including the clothing industry.

The increase in the inflow of FDI is a top priority of the Albanian government. In July 2013, after close consultations with the World Bank and the European Commission, the government approved the new seven years strategy on Business and Investment in Albania. The Minister of Economy, Trade, and Energy publicly stated that "by successfully implementing this strategy the government aims to attract until 2020 a FDI inflow amounting to 1.2 billion EUR". In addition, international organizations lead by the World Bank are in favor to support implementation of this strategy. The Chief Economist of the Europe and Central Asia Region of the World Bank[5] stated, "it is the time for Albania to move from a consumption-oriented economy towards a production-oriented economy. Foreign investments can help in a smooth transition".

This section is dedicated to find determinants of Albania, one of the countries in the WB region for further integration into European GVCs (see Fig. 3.5). Albania, as one of the members of the Western Balkans countries and as such it is mainly integrated into the WB regional value chain and into the European Union GVC.

In order to examine the level of integration of Albania in GVCs, initially key macro economic indicators taken from World Integrated Solutions are presented. The indicators range from 1997 up to 2011.[6] Referring to Fig. 3.6 direct labor value added is more than 70% of the total labor value added in the exports of Albania. This indicates that only a small proportion of the total value goes for manufacturing of intermediate products that are used in exports of final goods. This indicates that backward linkages prevail in the Albanian economy.

Figure 3.7 points out that the total value added of skilled labor in generating exports of Albania is one third of the value added of unskilled labor. This indicates

---

[4]United Nations prepared the two reports for the presence of FDI only for two consecutive years 2010 and 2011.

[5]Statement made on the conference "Albania A New Generation" that took place in Albania on August 27, 2013.

[6]These are the latest available observation for Albania in the World Integrated Solutions database.

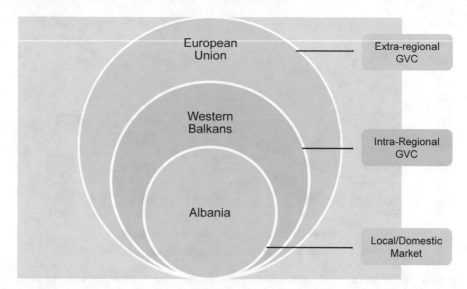

Fig. 3.5 The regional value chains in the Western Balkans and the European GVC. *Source* Author's drawing

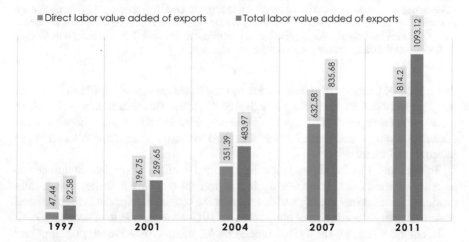

Fig. 3.6 Direct (Direct labor value added of exports the total (skilled plus unskilled) domestic labor value added embodied directly in a sector's exports, or the wages paid directly for the production of the sector's exports (measured in millions of nominal US$).) labor and total labor value added (The total (skilled plus unskilled) domestic labor value added embodied in a sector's exports, including the wages paid directly for the production of the sector's exports and indirectly via the production of economy-wide inputs for the sector's exports (backward linkages) (measured in millions of nominal US$).) of exports in Albania 2007–2011 (Mln/US$). *Source* World Integrated Solutions

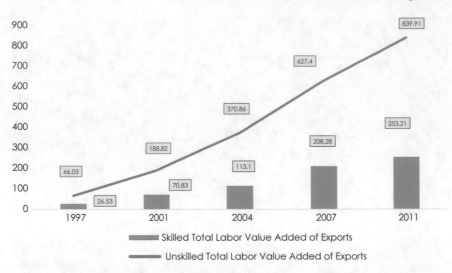

Fig. 3.7 Skilled total labor (The skilled domestic labor value added embodied in a sector's exports, including the skilled wages paid directly for the production of the sector's exports and indirectly via the production of economy-wide inputs for the sector's exports (backward linkages) (measured in millions of nominal US$).) and unskilled total labor value added (The unskilled domestic labor value added embodied in a sector's exports, including the unskilled wages paid directly for the production of the sector's exports and indirectly via the production of economy-wide inputs for the sector's exports (backward linkages) (measured in millions of nominal US$).) of exports in Albania 2007–2011 (Mln/US$). *Source* World Integrated Solutions

that up to 2011 exports of Albania were mostly based on low value added functions like assembly that require only a low degree of skills. This also indicates that there is a significant gap with the level of high value added exports in the generated in the Albanian economy and the existence of limited forward linkages that Albania has in international trade.

To continue, Fig. 3.8 shows that a very limited quantity of outputs manufactured in Albania are used as intermediary inputs in exports. This indicates that locally produced goods do not qualify and do not meet the standards required in international trade indicating a low degree of integration in GVCs especially in the European GVC.

In addition, Fig. 3.9 shows that exports of Albania are based mostly on imported goods, which are re-exported as final goods to end customers. This inward processing regime of exports is also reflected in the indirect value added of exports that are half of the direct value, indicating that Albania has still a lengthy road for participation in GVCs.

In order to identify the determinants that might affect the integration of Albania in GVCs the EORA database, the World Bank database and period data from the Bank of Albania and INSTAT are used to construct our time series dataset for Albania. The data set is used based on yearly observations ranging from 1990 to 2018. The dependent variable is GVC participation, which is retrieved from the EORA Multi Region Input-Output database.

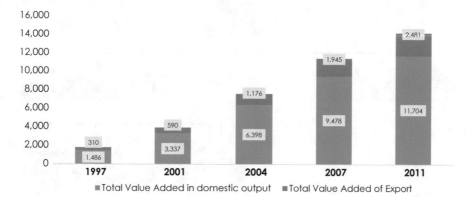

Fig. 3.8 Evolution of total value added in domestic output (The total domestic value added of a sector embodied in economy-wide production, including the direct value added and the indirect value added contained as inputs (forwards linkages) (measured in US$ mil).) and total value added of exports (The total domestic value added of a sector embodied in economy-wide exports, including the direct value added and the indirect value added contained as inputs (forwards linkages) (measured in US$ mil).) in Albania 2007–2011 (Mln/US$). *Source* World Integrated Solutions

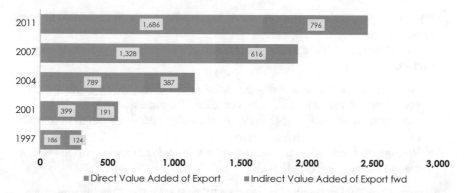

Fig. 3.9 Evolution of direct value added (The sector's domestic value added embodied in its exports, measured as gross exports less domestic and foreign inputs (measured in US$ mil).) and indirect value added in exports (The domestic value added of a sector embodied as inputs in economy-wide exports (forward linkages) (measured in US$ mil).) in Albania 2007–2011 (Mln/US$). *Source* World Integrated Solutions

GVC participation measures the participation of each sector j in a given country n in the cross-national trade of intermediate goods and it is calculated based on the method proposed by Koopman et al. [66]:

$$\text{GVC PARTICIPATION}_{jn} = \text{FVA}_{jn} + \text{IVA}_{jn},$$

where FVA (foreign value added) is a downstream stage of GVC, because it is the content of intermediate imports incorporated in gross exports, and IVA (indirect value

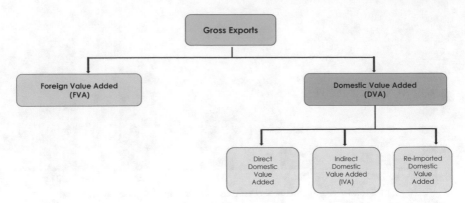

Fig. 3.10 Decomposition of gross exports. *Source* Author's adaption based on Koopman et al. [66]

added) is an upstream stage of GVC, because it is the indirect domestic value added to both sector j and country n, divided by total sector-country exports [3].

In order to better explain the above definition of Koopman et al. [66] he decomposes gross exports into two main components (see Fig. 3.10):

1. FVA (foreign value added)—content of intermediate imports embodied in gross exports
2. DVA (domestic value added)—the value of domestically produced exports divided in:

   - Direct domestic value added—the value added embodied in exports of final goods and intermediates, absorbed by direct importers;
   - Indirect domestic value added (IVA)—the value added embodied in intermediates re-exported to third countries;
   - Re-imported domestic value added—the value added from exported intermediates that are reimported.

In order to capture the percentage change (elasticity) rather than the unit change, GVC Participation (measured in current USD) is transformed to its logarithmic value. Based on the literature review and on the data availability for Albania, the following indicators are used as independent variables (see Table 3.2).

Before estimating the models, descriptive statistics based on the data sets of variables under consideration are analyzed. The descriptive statistics are presented in Table 3.7 in the appendix of this chapter. With regard to the model the analysis is based on the OLS (Ordinary Least Squares) having the following equation:

$$Y_t = \alpha + \beta X_t + \varepsilon_t$$

- Y_t refers to the dependent variable at time t,
- α is the intercept,
- X_t is the vector of independent variables at time t,

Table 3.2 Independent variables, description and the expected impact sign

Independent variables name	Description	Explanation/source	Expected impact sign
GDPg	GDP growth (annual %) is an indicator of the level of development and economic potential	Škabić [94]. *Source* World Bank	+
lGDP_cap	GDP per capita (USD) is an indicator of the level of development and economic potential	In order to capture the percentage change (elasticity) rather than the unit change, the GDP per capita (measured in current USD) is transformed to its logarithmic value. Škabić [94]. *Source* World Bank	+
lGNI_cap	GNI per Capita (USD) is an indicator of the level of development and economic potential	In order to capture the percentage change (elasticity) rather than the unit change, GNI per capita (measured in current USD) is transformed to its logarithmic value. *Source* World Bank	+
unemp	The unemployment rate is an indicator of the level of development and economic potential	It is a lagging indicator, meaning that it generally rises or falls in the wake of changing economic conditions, rather than anticipating them. *Source* World Bank	−
FDIstGDP	FDI stock to GDP ratio	FDI has been identified as the most common way to link developing countries to GVCs. Taglioni and Winkler [95], Škabić [94]. *Source* Bank of Albania	+

(continued)

Table 3.2 (continued)

Independent variables name	Description	Explanation/source	Expected impact sign
FDIinfGDP	FDI inflow to GDP ratio	FDI has been identified as the most common way to link developing countries to GVCs. Taglioni and Winkler [95], Škabić [94]. *Source* Bank of Albania	+
lNMW	National min wage (USD) is an indicator of the location advantage for attraction of foreign direct investments	In order to capture the percentage change (elasticity) rather than the unit change, NMW (measured in current USD) is transformed to its logarithmic value. Škabić [94]. *Source* World Bank	± (The negative relation is explained by the fact that wages lead to lower GVC participation in case of developing countries dominated from a low skilled labor force, while in developed countries that have specialized in high-technology and human capital higher wages will lead to higher GVC participation)
EdExp	Expenditure on Education as % of GDP considered as a proxy for the level of absorptive capacity needed to satisfy investors' demand for sophisticated intermediates	Borenszein et al. [12] and Amendolagine et al. [3]	±
EXP_GDP	Exports to GDP ratio	Exports to GDP ratio are an indicator of the importance of exports in the country's economy. *Source* INSTAT, Author's own calculations	+

(continued)

Table 3.2 (continued)

Independent variables name	Description	Explanation/source	Expected impact sign
PT	Profit tax is the amount of taxes on profits paid by the business. Profit tax % is important as a determinant of FDI attractiveness	Škabić [94]. *Source* World Bank	− (Higher profit taxes imply that a country is less attractive to FDI, resulting in lower result GVC participation)
DC	Domestic credit provided by financial sector (% of GDP). Financial development is important as the financial system is a provider of capital for businesses	Domestic credit provided by the financial sector includes all credit to various sectors on a gross basis, with the exception of credit to central government, which is net. The financial sector includes monetary authorities and deposit money banks, as well as other financial corporations where data are available (including corporations that do not accept transferable deposits but do incur such liabilities as time and savings deposits). Examples of other financial corporations are finance and leasing companies, money lenders, insurance corporations, pension funds, and foreign exchange companies. Škabić [94]. *Source* World Bank	+

- ε_t is the error term. In order to achieve the best-fitted model the following steps are performed.

3.3.2.1 Stationarity and the Dickey Fuller Test

To start with the series of the selected model need to be stationary. Time series are considered stationary when statistical properties such as mean, variance, autocorrelation are constant over time. Most of statistical forecasting methods are based on the assumption that time series can be rendered approximately stationary (i.e., "stationarized") through the use of mathematical transformations. The Dicky-Fuller test is one of the most popular tests to check the stationarity of the variables and it is based on the following hypothesis and decision criteria (see Table 3.3):

H_0: variable is non-stationary (e.g. there exists a unit root in the variable)
H_A: variable is stationary (e.g. there doesn't exist any unit root in the variable)
Decision criteria: Reject the null if MacKinnon p-value is less than 1%, 5%, or 10% level of significance.

Some of our variables such as GDP growth and unemployment rate are stationary at 5% and 1% respectively and there is no need to make further mathematical transformations. While, the rest of the variables become stationary after taking the first difference. Figures 3.11 and 3.12 are a sample of the transformation of variables

Table 3.3 Results of the Dickey Fuller test at a country level for participation in GVCs

Variable name	MacKinnon *p*-value	MacKinnon *p*-value at first difference
lGVC	0.7843	0.000***
GDPg	0.0106**	
lGDP_cap	0.9626	0.000***
lGNI_cap	0.9038	0.0041***
unemp	0.0023***	
FDIstGDP	0.9882	0.000***
FDIinfGDP	0.5948	0.000***
lNMW	0.7137	0.000***
EdExp	0.3821	0.0006***
EXP_GDP	0.7201	0.000***
PT	0.4012	0.0236**
DC	0.3516	0.0004***

Significant at ***1%, **5%, *10%
Statistical significance indicates the probability that the difference or observed relationship between a dependent and control variables is not by chance. The significance level of 1% indicates that the dependent and the control variables are highly related.

Fig. 3.11 Original series of
GVC participation. *Source*
Author's calculations

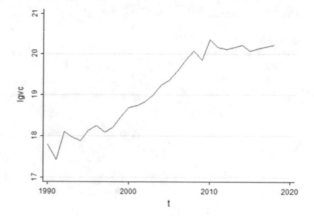

Fig. 3.12 First difference of
logarithmic transformation
of GVC participation. *Source*
Author's calculations

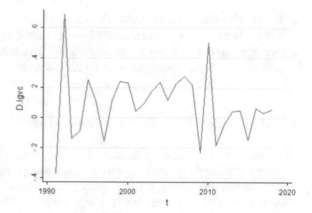

from non-stationary to stationary. The remaining transformations are included in the
appendix of this chapter (see Figs. 3.13, 3.14, 3.15, 3.16, 3.17, 3.18, 3.19, 3.20, 3.21,
3.22, 3.23, 3.24, 3.25, 3.26, 3.27, 3.28, 3.29, 3.30, 3.31, 3.32, 3.33, 3.34 and 3.35).

3.3.2.2 Multicollinearity and Correlation Matrix

Multicollinearity is another property examined in time series. It is examined through
the correlation matrix between the independent variables and a decision is made
based on the rule that if correlation is 0.8 or higher in absolute value it is indicative
for imperfect multicollinearity. Thus, these variables are not used in the same regression model. In addition, the check on the variance inflation factor is considered as a
useful way to look for multicollinearity amongst independent variables. The correlation matrix of selected independent variables in the appendix of this chapter (see
Table 3.8). Based on it's results the imperfect multicollinearity is addressed in one
of the following ways:

Fig. 3.13 Scatterplot
showing the relation between
GVC participation and GDP
growth. *Source* Author's
calculations

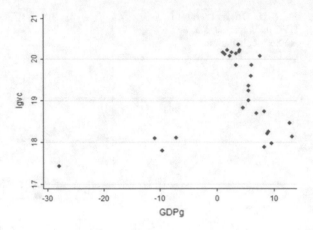

- The variable that is insignificant or theoretically invalid it is dropped. If the variable it is likely to cause omitted variable bias, it is not dropped.
- Consideration of additional transformations of variables such as logarithms, first differencing, lags, generating squared variables etc.

3.3.2.3   Selection of the Correct Functional Model Form

The selection of the correct functional model form is based upon examining the scatterplots of the dependent variable and each independent variable. This is done in order to check if there are any non-linear relationships. The majority of scatterplots show a linear relationship between the dependent variable lGVC (GVC participation) and independent variables (see Fig. 3.12). The only non-linear relationships that are between GVC participation and GDP growth, GVC participation and unemployment rate, GVC participation and FDI stock as a ratio to GDP, GVC participation and Profit Tax. Following this analysis the square of these independent variables is considered, based on the assumption that the model might be non-linear. However, none of the squared variables showed significance, as a result they are not included in the model. Below is presented the scatterplot between the GVC participation and the GDP growth. The remaining scatterplots are presented in the appendix of this chapter (see Figs. 3.36, 3.37, 3.38, 3.39, 3.40, 3.41, 3.42, 3.43 and 3.44).

3.3.2.4   Model Linearity and Ramsey Regression Equation
             Specialization Error Test (RESET)

In statistics, the Ramsey Regression Equation Specification Error Test (RESET) test is a general specification test for the linear regression model. More specifically, it tests whether non-linear combinations of the fitted values help explain the dependent variable. The intuition behind the test is that if non-linear combinations of explanatory (independent) variables have any power in explaining the response (dependent) variable. If this is the case then the model is miss-specified, in the sense that the data generating process might be better approximated by a non-linear functional form.

H_0: The model has no omitted variables
H_A: The model has omitted variables
Decision criteria: Reject the null if p-value is less than 1%, 5%, or 10% level of significance. Rejection of the null hypothesis implies that there are possible missing variables and biased estimators caused by the presence of endogeneity. In economic models endogeneity broadly refers to situations in which an explanatory variable is correlated with the error term.

3.3.2.5   Autocorrelation and the Breusch-Godfrey Langrage Multiplier Test

The Breusch–Godfrey serial correlation LM test is a test for autocorrelation in the errors of the regression model. Serial correlation is defined as the correlation between the observations of residuals and it is commonly found in time series data. It makes use of the residuals from the model, which are considered in a regression analysis.

H_0: there is no serial correlation
H_A: there is serial correlation
Decision criteria: Reject the null if p-value is less than 1%, 5%, or 10% level of significance.

3.3.2.6   Heteroscedasticity-Breusch-Pagan Test

The Breusch-Pagan Test begins by allowing the hetcroskedasticity process to be a function of one or more independent variables, and it's usually applied by assuming that heteroskedasticity may be a linear function of all independent variables in the model.

H_0: there exists homoscedasticity (e.g. error variances are all equal)
H_A: there exists heteroscedasticity (e.g. error variances are a multiplicative function of one or more variables)
Decision criteria: Reject the null if p-value is less than 1%, 5%, or 10% level of significance.

All the above listed tests are applied in the analysis and after tens of trials the results of the OLS (Ordinary Least Squares) estimations of the five best models are presented. The best models based on the highest R-squared (a statistical measure that represents the proportion of the variance for a dependent variable), the fulfillment of the qualifying criteria for the tests previously mentioned and the significance of the coefficients. The five most appropriated models are:

1. $lGVC = \alpha + \beta_1 lGDP\_cap + \beta_2 unemp + \beta_3 FDIstGDP + \beta_4 EdExp + \beta_4 Exp\_GDP + \beta_5 PT + \beta_6 DC + \varepsilon$
2. $lGVC = \alpha + \beta_1 lGDP_{cap} + \beta_2 FDIinfGDP + \beta_3 lNMW + \beta_4 Exp_{GDP} + \varepsilon$
3. $lGVC = \alpha + \beta_1 FDIinfGDP + \beta_2 lNMW + \beta_3 Exp\_GDP + \varepsilon$
4. $lGVC = \alpha + \beta_1 L1.lGVC + \beta_2 lGDP\_cap + \beta_3 unemp + \beta_4 FDIstGDP + \beta_4 EdExp + \beta_5 Exp\_GDP + \varepsilon_t$
5. $lGVC = \alpha + \beta_1 lGDP\_cap + \beta_2 FDIinfGDP + \beta_3 lNMW + \beta_4 EdExp + \beta_4 Exp\_GDP + \beta_5 DC + \varepsilon_t$

3.3.2.7 Results

The initial results of OLS estimations of the determinants of Albania's participation in GVCs are shown in Table 3.4.

The results confirm that Albania is not well integrated in GVCs and there is need to attract additional FDIs in order to improve the country's participation in GVCs. Some results are in line with the existing literature. Similarly, to Škabić [94] the results indicate that the ratio of FDI inflows to GDP has a positive impact on GVC participation. Therefore, FDI inflows to GDP are a key determinant for participation in GVCs, especially for developing countries such as Albania.

Another finding is that the logarithm of the national minimum wage of Albania has a positive significant effect on GVC participation. This result is counterintuitive,

Table 3.4 Initial results from OLS estimations for five best models on the participation of Albania in GVCs

Independent variable	Model 1	Model 2	Model 3	Model 4	Model 5
L1.lGVC				0.3859144*	
GDPg		0.0007194			
lGDP_cap	7.926721**				−0.4351520
lGNI_cap				0.9181388*	
unemp	0.0323034			−0.0250763	
FDIstGDP	−3.92723**			−1.120494	
FDIinfGDP		2.65113	2.585497		2.542783
lNMW		0.5234458***	0.510921***		0.6925648**
EdExp	0.1365444			−0.1685778	−0.1133632
EXP_GDP	9.465017**	3.379777**	3.801808***	4.019638**	5.435358**
PT	0.0220515				
DC	−0.0213981				0.0099185
constant	−0.7161041	−0.0130397	0.0135468	0.3817866	0.0026767
R-squared	0.8310	0.5185	0.5319	0.6291	0.6407

Significant at ***1%, **5%, *10%

because in developing countries, lower wages attract more FDIs. This is in line with the data generated from World Integrated Trade Solutions as exports in Albania are mostly low value added and generated from a low skilled labor force.

In addition, GDP per capita has a positive and significant impact in GVC participation, which is in line with Škabić [94]. On the other hand, GDP growth was insignificant in affecting GVC participation. These two variables show the level of development and economic potential of the country. The more developed the country, the more it will be integrated in GVC participation. Moreover, Profit Tax is also insignificant in affecting Albania's GVC participation at the country level.

This results of the initial tests are indicative of the existence of autocorrelation. As such, the Breusch-Godfrey Langrage Multiplier Test for autocorrelation is run for each of the above mentioned models. The null hypothesis is that there is no serial correlation is rejected at 5% level of significance. Table 3.5 shows the results from the OLS estimations of our five best models after serial correlation is taken into account.

Given the results of initials models the results indicate that after serial correlation is addressed the R-squared increases, as well as the number of significant variables. The variable of financial development, which is measured by domestic credit as percentage to GDP, has a positive significant effect on GVC participation after serial correlation is taken into account. Therefore, the selected best model is the following:

$$lGVC = 3.593657FDI\,inf\,GDP + 0.6071325lNMW - 0.1858721EdExp$$
$$+ 5.221439Exp\_GDP + 0.0117944DC + \varepsilon_t$$

Table 3.5 Final results from OLS estimations for our five best models after curing for serial correlation

Independent variable	Model 1	Model 2	Model 3	Model 4	Model 5
L1.lGVC				0.146868	
GDPgsq		0.0004135			
lGDP_cap	7.393024**				0.2803419
lGNI_cap				0.9184136**	
unemp	0.0415261			−0.0082695	
FDIstGDP	−3.525057**			−0.3373567	
FDIinfGDP		0.0004035*	3.058731*		3.396679**
lNMW		0.4617345***	0.4559453***		0.5656601***
EdExp	0.1872582			−0.1314903	−0.1861685*
EXP_GDP	9.065399***	3.281623**	3.561589***	3.656132**	4.882326***
PT	0.0106543				
DC	−0.0067251				0.011954***
Constant	−0.783858	−0.0009356	0.0163883	0.1621991	−0.0140874
R-squared	0.8970	0.6361	0.5310	0.7231	0.8772
Rho	−0.7071479	−0.4593128	−0.4478835	−0.584202	−0.7681473
Durbin-Watson statistics	3.119299	2.804835	2.887060	2.381021	3.326805

Significant at ***1%, **5%, *10%

The results show that FDI inflows as a ratio to GDP contribute positively to GVC participation, emphasizing the fact that FDI in an important determinant for a country's integration in GVCs. The nominal minimum wage also positively affects GVC participation, which in the case of developing countries such as Albania is counterintuitive. Education expenditure as percentage of GDP has a negative significant effect on GVC participation. This result can be explained by the fact that developing countries are often exploited for their cheap labor force and human capital. Exports to GDP ratio shows the importance of exports in the economic activity of a country and therefore higher exports to GDP ratio determines higher GVC participation. Last, higher levels of domestic credit as percentage to GDP contribute to higher GVC participation as an easy credit serves as facilitating instrument for further investments.

3.3.2.8   Limitations

The models applied have their limitations regarding the number of observations and the available variables to be included in the analysis. The data collected for Albania has a limited number of observations as the country turned into an open market economy only in the early 1990s. Statistical tests give more accurate results when the dataset is large. The study can be expanded by analyzing a panel data set of Albanian firms by including independent variables such as variable measured as the log of years since the first investment [3], variable showing the size and the age of the firm, variable showing the percentage owned by foreign investors etc.

Also, there are still a number of variables used in other studies that are not estimated for Albania and for the countries in the WB region. Such variables are included in Table 3.6.

Table 3.6 Explanatory variables that can be further included in model estimation

Variable name	Description	Source
R&D	Investment in Research and Development as % of GDP as an indicator for innovation. The assumption is that with higher level of investment in R&D more sophisticated production will arise, and from the existing research such sectors are more involved in GVCs	Taglioni and Winkler [95], Škabić [94]
PRF (Property Rights Freedom)	This variable can be used to measure the foreign investors' guarantee of their rights/assets	Škabić [94]

Appendix

See Tables 3.7, 3.8 and Figs. 3.14, 3.15, 3.16, 3.17, 3.18, 3.19, 3.20, 3.21, 3.22, 3.23, 3.24, 3.25, 3.26, 3.27, 3.28, 3.29, 3.30, 3.31, 3.32, 3.33, 3.34, 3.35, 3.36, 3.37, 3.38, 3.39, 3.40, 3.41, 3.42, 3.43 and 3.44.

Table 3.7 Descriptive statistics of dependent and independent variables in GVC participation at a country level for Albania

Variable name	Observations	Mean	Standard deviation	Min.	Max.
lgvc	29	19.14469	0.9577716	17.41618	20.35414
lgdp_cap	29	7.918502	0.4482106	7.125,283	8.532082
lgni_cap	29	7.451958	0.9528193	5.634789	8.488793
lnmw	24	4.60257	0.7120968	3.273364	5.459586
gdpg	29	3.078021	8.188006	−28.00214	13.32233
gdp_cap	29	3010.448	1238.721	1243	5075
gni_cap	29	2454.828	1720.871	280	4860
unemp	29	15.29961	3.437018	9.1	26.5
fdistgdp	27	0.2116256	0.1427837	0.0306666	0.5247372
fdiinfgdp	27	0.0547021	0.0270158	0.0128264	0.0990546
edexp	22	3.489545	0.6208173	2.3	4.5
exp_gdp	29	0.2142772	0.0733633	0.0748482	0.3172197
pt	14	13.02857	4.189259	8.1	20.1
dc	25	57.05083	9.047237	38.83821	69.51783

Table 3.8 Correlation matrix of independent variables possibly contributing to GVC participation at country level for Albania

	D. lgdp_cap	D. lgni_cap	D. lnmw	gdpg	unemp	D. fdistgdp	D. fdiinf-p	D. edexp	D. exp_gdp	D. pt	D. dc
D1.lgdp_cap	1.0000										
D1.lgni_cap	0.8447	1.0000									
D1.lnmw	0.3146	0.2705	1.0000								
gdpg	0.9972	0.8248	0.2953	1.0000							
unemp	−0.3734	−0.3125	−0.0668	−0.3800	1.0000						
D1.fdistgdp	0.1070	−0.1662	−0.5353	0.1161	0.0603	1.0000					
D1.fdiinfgdp	0.3037	0.4670	0.1800	0.2852	0.0634	−0.3305	1.0000				
D1.edexp	−0.0425	0.0882	0.1572	−0.0514	−0.6179	−0.1817	0.1320	1.0000			
D1.exp_gdp	0.0844	−0.0296	−0.2793	0.0916	−0.0705	0.5707	0.0273	−0.1905	1.0000		
D1.pt	−0.2761	−0.3844	0.0838	−0.2389	0.4568	−0.0418	−0.1242	−0.2114	−0.0167	1.0000	
D1.dc	0.6823	0.8747	−0.0212	0.6574	−0.2733	0.0102	0.2378	−0.0466	0.0781	−0.4528	1.0000

Fig. 3.14 Original time
series of log GVC
participation

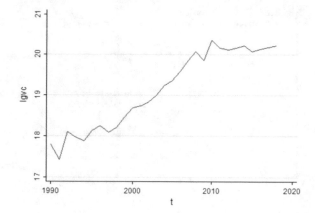

Fig. 3.15 First difference of
log GVC participation series

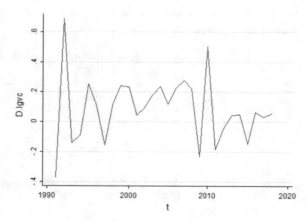

Fig. 3.16 Original time
series of log of GDP per
capita

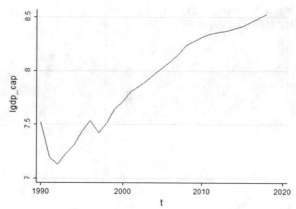

Fig. 3.17 First difference of
log of GDP per capita series

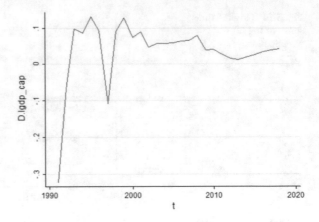

Fig. 3.18 Original time
series of log of GNI per
capita

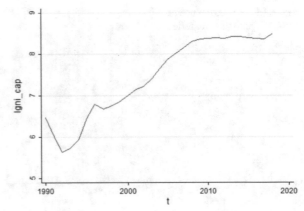

Fig. 3.19 First difference of
log of GNI per capita series

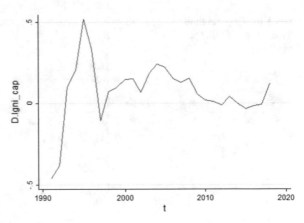

Fig. 3.20 Original time
series of log of nominal
minimum wage

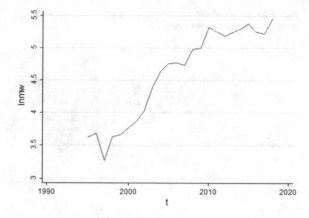

Fig. 3.21 First difference of
log of national minimum
wage series

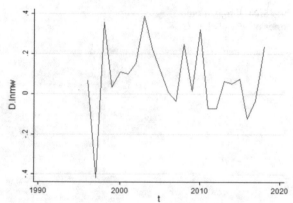

Fig. 3.22 Original time
series of GDP growth

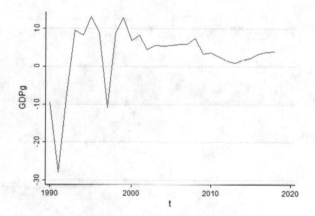

Fig. 3.23 Original time
series of unemployment rate

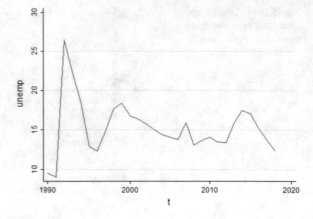

Fig. 3.24 Original time
series of FDI stock to GDP

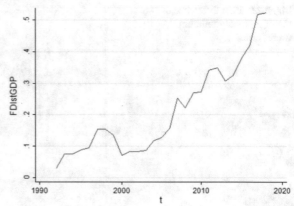

Fig. 3.25 First difference of
FDI stock to GDP series

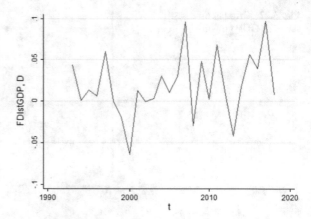

Fig. 3.26 Original time series of FDI Inflow to GDP

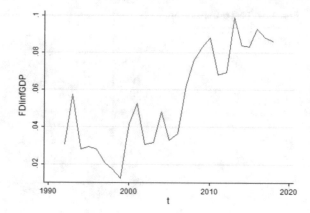

Fig. 3.27 First difference of FDI inflow to GDP series

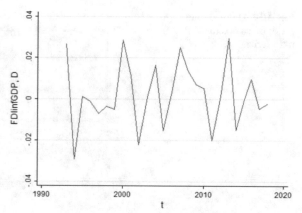

Fig. 3.28 Original time series of education expenditure to GDP

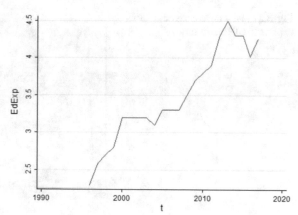

Fig. 3.29 First difference of
education expenditure to
GDP series

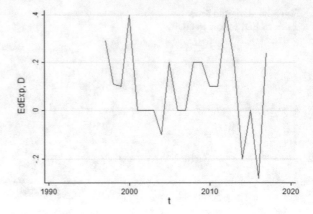

Fig. 3.30 Original time
series of exports to GDP
ratio series

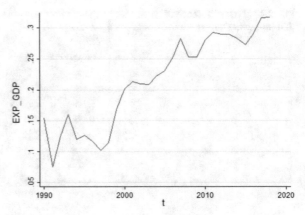

Fig. 3.31 First difference of
exports to GDP ratio series

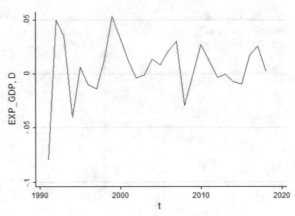

Fig. 3.32 Original series of profit tax as % to GDP

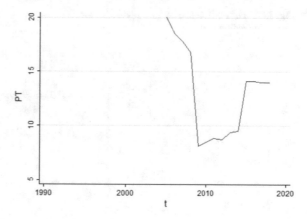

Fig. 3.33 First difference of profit tax as % to GDP series

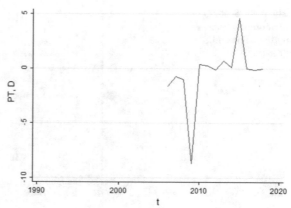

Fig. 3.34 Original time series of domestic credit

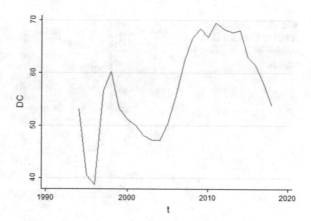

Fig. 3.35 First difference of
domestic credit series

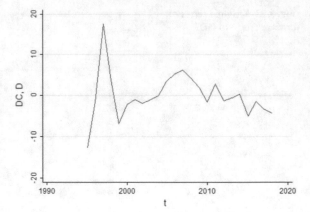

Fig. 3.36 Scatterplot
relation between GVC
participation and GDP
growth

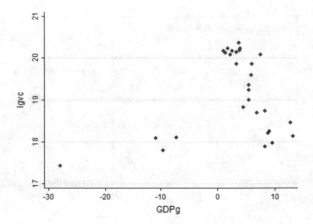

Fig. 3.37 Scatterplot
relation between GVC
participation and GNI per
capita

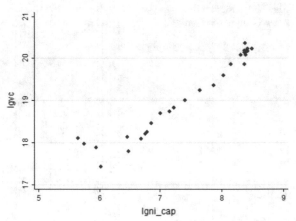

Fig. 3.38 Scatterplot relation between GVC participation and GDP per capita

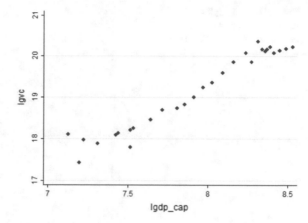

Fig. 3.39 Scatterplot relation between GVC participation and FDI stock to GDP

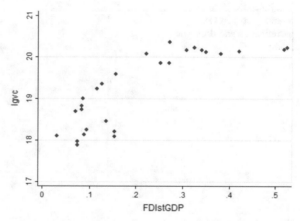

Fig. 3.40 Scatterplot relation between GVC participation and exports to GDP

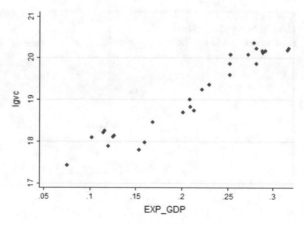

Fig. 3.41 Scatterplot
relation between GVC
participation and profit tax as
a % to GDP

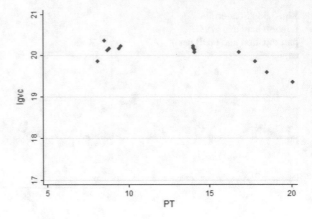

Fig. 3.42 Scatterplot
relation between GVC
participation and domestic
credit

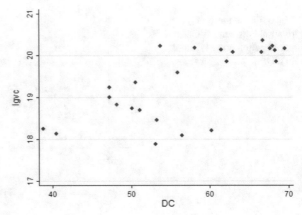

Fig. 3.43 Scatterplot
relation between GVC
participation and education
expenditure as a % to GDP

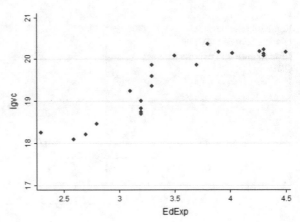

Fig. 3.44 Scatterplot relation between GVC participation and unemployment rate

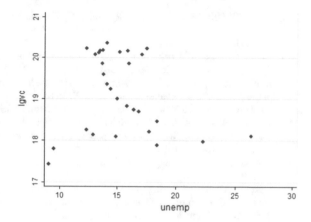

References

1. Acemoglu D (2019) Automation and new tasks: how technology displaces and reinstates labor? J Econ Perspect 33(2):3–30
2. Acemoglu D, Restrepo P (2018) Low-Skill and high-skill automation. J Hum Capital 12(2):204–232
3. Amendolagine V, Presbitero AF, Rabellotti R, Sanfilippo M, Seric A (2017) FDI, global value chains, and local sourcing in developing countries. IMF working paper WP/17/284, Washington D.C.
4. Antràs P, De Gortari A (2017) On the geography of global value chains. NBER Working Paper 23456. National Bureau of Economic Research, Cambridge, MA
5. Atoyan RV, Benedek D, Cabezon E, Cipollone G, Miniane JA, Nguyen N, Petri M, Reinke J, Roaf J (2018) Public infrastructure in the Western Balkans: opportunities and challenges. IMF Publications, Washington D.C.
6. Bas M, Strauss-Kahn V (2014) Does importing more inputs raise exports? Firm-level evidence from France. Rev World Econ 150:241–275
7. Bas M, Strauss-Kahn V (2015) Input-trade liberalization, export prices, and quality upgrading. J Int Econ 95(2):250–262
8. Belderbos R, Sleuwaegen L, Somers D, De Backer K (2016) Where to locate innovative activities in global value chains: does co-location matter? In: OECD science, technology and industry policy papers 30. OECD Publishing, Paris
9. Bernard AB, Moxnes A, Ulltveit-Moe KH (2018) Two-sided heterogeneity and trade. Rev Econ Stat 100(3):424–439
10. Bhatia US (2013) The globalization of supply chains—policy challenges for developing countries. In: Elms DK, Low P (eds) Global value chains in a changing world, pp 195–220. Retrieved from https://www.wto.org/english/res_e/booksp_e/aid4tradeglobalvalue13_e.pdf
11. Bilir LK, Chor D, Manova K (2016) Host-country financial development and multinational activity. Eur Econ Rev 115:192–220
12. Borensztein E, Gregorio JD, Lee J-W (1998) How does foreign direct investment affect economic growth? J Int Econ 45(1):115–135
13. Borin A, Mancini M (2019) Measuring what matters in global value chains and value-added trade. Policy research working paper. No. WPS 8804, WDR 2020 Background paper. World Bank Group, Washington D.C.
14. Bown CP, Freund CL (2019) Active labor market policies: lessons from other countries for the United States. PIIE working paper 19–2. Peterson Institute for International Economics, Washington, D.C.

15. Brown D, Dehejia R, Robertson R, Domat G, Veeraragoo S (2015) Are sweatshops profit-maximizing? Answer: No; evidence from better work Vietnam. Better Work Discussion Paper, International Labour Office, Geneva
16. Bucini G, Finotto V (2016) Innovation in global value chains: co-location of production development in Italian low-tech industries. Reg Stud 50(12):2010–2023
17. Buelens C, Tirpak M (2017) Reading the footprints: how foreign investors shape countries' participation in global value chains. Comp Econ Stud 59(4):561–584
18. Civici A (2012) 100 Years of Albanian economy. Monit Mag 556:11–15
19. Coelli F, Moxnes A, Ulltveit-Moe KH (2016) Better, faster, stronger: global innovation and trade liberalization. NBER working paper 22647. National Bureau of Economic Research, Cambridge, Massachusetts
20. Costinot A, Vogel J, Wang S (2013) An elementary theory of global supply chains. Rev Econ Stud 80(1):109–144
21. Cusolito AP, Safadi R, Taglioni D (2016) Inclusive global value chains: policy options for small and medium enterprises and low-income countries. Directions in development: trade series. World Bank and OECD, Washington, D.C.
22. Dassbach CHA (1989) Global enterprises and the world economy: Ford, GM and IBM, the emergence of the transnational enterprise. Garland, New York
23. Davies W (2012) The emerging neocommunitarianism. Polit Q 83:767–776
24. De Backer K, Miroudot S (2013) Mapping global value chains. OECD Trade Policy Papers, No. 159. OECD Publishing, Paris
25. De Backer K, Destefano T, Moussiegt L (2017) The links between global value chains and global innovation networks: an exploration. OECD science, technology and innovation policy papers, No. 37. OECD Publishing, Paris
26. De Gortari A (2019) Disentangling global value chains. NBER working paper 25868. National Bureau of Economic Research, Cambridge, Massachusets
27. De Lucia A (2006) Albanian cultural profile, european commission project, cultural sensitivity and competence in adolescent mental health promotion, prevention and early intervention (CSCAMHPPEI-015127). Sixth framework programme: specific support action, dipartimento di scienze statistiche, università degli studi di bari, Italia
28. Defever F, Heid B, Larch M (2015) Spatial exporters. J Int Econ 95(1):145–156
29. Del Prete D, Giovannetti G, Marvasi E (2017) Global value chains participation and productivity gains for north african countries. Rev World Econ 153(4):675–701
30. Duggan V, Rahardja S, Varela GJ (2015) Revealing the impact of relaxing service sector FDI restrictions on productivity in Indonesian manufacturing, policy note 5. World Bank, Jakarta, Indonesia
31. Farole T, Winkler D (2014) Making foreign direct investment work for Sub-Saharan Africa. Local spillovers and competitiveness in global value chains. The World Bank, Washington D.C.
32. Farole T, Winkler D (2015) The role of foreign firm characteristics, absorptive capacity and the institutional framework for FDI spillovers. J Bank Financ Econ 1(3):77–112
33. Fernandes AM, Freund CL, Pierola MD (2016) Exporter behavior, country size and stage of development: evidence from the exporter dynamics database. J Dev Econ 119:121–137
34. Fernandes AM, Hillberry RH, Mendoza-Alcántara A (2019a) Trade effects of customs reform: evidence from Albania. Paper presented at department of economics spring seminar. University of Nebraska, Lincoln
35. Fernandes AM, Kee HL, Winkler D (2019b) Factors affecting global value chain participation across countries. Policy Research Working Paper. World Bank, Washington D.C.
36. Fessehaie J (2017) Leveraging the services sector for inclusive value chains in developing countries. In: International centre for trade and sustainable development (ICTSD), Geneva, Switzerland
37. Fort TC (2017) Technology and production fragmentation: domestic versus foreign sourcing. Rev Econ Stud 84(2):650–687

38. Freund CL, Moran TH (2017) Multinational investors as export superstars: how emerging-market governments can reshape comparative advantage. Working Paper 17–1, Peterson Institute for International Economics, Washington D.C.
39. Frick SA, Rodríguez-Pose A, Wong MD (2019) Toward economically dynamic special economic zones in emerging countries. Econ Geogr 95(1):30–64
40. Froebel F, Heinrich J, Kreye G (1980) The new international division of labour: structural unemployment in industrialised countries and industrialisation in developing countries. Cambridge University Press, Cambridge
41. Gelb A, Meyer CJ, Ramachandran V, Wadhwa D (2017) Can Africa be a manufacturing destination? Labor costs in comparative perspective. CGD working paper 466. Center for Global Development, Washington D.C.
42. Gereffi G (1994) The organization of buyer-driven global commodity chains: how US retailers shape overseas production networks. Commodity Chains and Global Capitalism, Praeger, pp 95–122
43. Gereffi G (2001) Shifting Governance structures in global commodity chains, with special reference to the internet. Am Behav Sci 44(10):1616–1637
44. Gereffi G, Lee J (2012) Why the world suddenly cares about global supply chains. J Supply Chain Manag 43(3):24–32
45. Gereffi G, Memedovic O (2003) The Global apparel value chain: what prospects for upgrading by developing countries?. United Nations Industrial Development Organization, Vienna
46. Gereffi G, Fernandez-Stark (2011) Global value chain analysis: a primer. Duke University, Global Value Chains (GVC) Center
47. Gereffi G, Sturgeon T (2013) Global value chain—oriented industrial policy: the role of emerging economies. In: Elms DK, Low P (eds) Global value chains in a changing world, pp. 195–220. Retrieved from https://www.wto.org/english/res_e/booksp_e/aid4tradeglobalvalue13_e.pdf
48. Glas A, Hübler M, Nunnenkamp P (2016) Catching up of emerging economies: the role of capital goods imports, FDI inflows, domestic investment and absorptive capacity. Appl Econ Lett 23(2):117–120
49. Godfrey S (2015) Global, regional and domestic apparel value chains in Southern Africa: social upgrading for some and downgrading for others. Cambridge J Reg Econ Soc 8(3):491–504
50. Hakro NA, Fida AB (2009) Trade and income convergence in selected South Asian countries and their trading partners. Lahore J Econ 14(2):49–70
51. Hallward-Driemeier M, Nayyar G (2018) Trouble in the Making? The future of manufacturing-led development. World Bank, Washington D.C.
52. Halpern L, Koren M, Szeidl A (2015) Imported inputs and productivity. Am Econ Rev 105(12):3660–3703
53. Henn C, Papageorgiou C, Romero JM, Spatafora N (2017) Export quality in advanced and developing economies: evidence from a new data set. Policy research working paper 8196. World Bank, Washington D.C.
54. Herzer D, Nowak-Lehmann F, Siliverstovs B (2006) Export-led growth in Chile: assessing the role of export composition in productivity growth. Dev Econ 44(3):306–328
55. Hopkins T, Wallerstein I (1977) Patterns of development of the modern world. Syst Rev 1(2):11–45
56. Hummels D, Schaur G (2013) Time as a trade barrier. Am Econ Rev 103:2935–2959
57. Humphrey J, Schmitz H (2002) How does insertion in global value chains affect upgrading in industrial clusters? Reg Stud 36(9):1017–1027
58. Ignatenko A, Raei F, Mircheva B (2019) Global value chains: what are the benefits and why do countries participate? IMF working paper WP/19/18, Washington D.C.
59. Ilahi N, Khachatryan A, Lindquist W, Nguyen N, Raei F, Rahman J (2019) Lifting growth in the Western Balkans. The role of global value chains and services exports. IMF working paper No. 19/13, Washington D.C.

60. Jirasavetakul LBF, Rahman J (2018) Foreign direct investment in new member states of the EU and Western Balkans: taking stock and assessing prospects. IMF working paper, Washington D.C.
61. Johnson RC (2018) Measuring global value chains. Ann Rev Econ 10(1):207–236
62. Kaplinsky R, Morris M (2002) A handbook for value chain. International Development Research Centre, Canada
63. Kee HL, Tang H (2016) Domestic value added in exports: theory and firm evidence from China. Am Econ Rev 106(6):1402–1436
64. Kee HL (2015) Local intermediate inputs and the shared supplier spillovers of foreign direct investment. J Dev Econ 112(C):56–71
65. Kimura F, Ando M (2010) Two-dimensional fragmentation in East Asia: conceptual framework and empirics. Int Rev Econ Finance 14(3):17–348
66. Koopman R, Powers W, Wang Zh, Wei Sh-J (2010) Give credit where credit is due: tracing value added in global production chains. Working Paper, No.16426, National Bureau of Economic Research
67. Koopman R, Wang Z, Wei SJ (2012) Tracing value-added and double counting in gross exports. NBER working paper No. 18579. National Bureau of Economic Research, Cambridge, MA
68. Kowalski P, Lopez-Gonzales J, Ragoussis A, Ugarte C (2015) Participation of developing countries in global value chains. OECD trade policy papers, No. 179
69. Kummritz V, Beverelli C, Koopman RB, Neumueller S (2016) Domestic foundations of global value chains. World Bank, Washington D.C.
70. Lam J, Postle R (2006) Textile and apparel supply chain management in Hong Kong. Int J Clothing Sci Technol 18(4):265–277
71. Lanz R, Piermartini R (2016) Comparative advantage in supply chains. World Trade Organization, Working Paper, Geneva
72. Lanz R, Piermartini R (2018) Specialization within global value chains: the role of additive transport costs. WTO staff working papers ERSD-2018-05. Economic research and statistics division. World Trade Organization, Geneva
73. Lessard D (2013) Uncertainty and risk in global supply chains. In: Elms DK, Low P (eds) Global value chains in a changing world, pp 195–220. Retrieved from https://www.wto.org/english/res_e/booksp_e/aid4tradeglobalvalue13_e.pdf
74. Lychain S, Pinkse J, Slade ME, Van Reen J (2016) Spillovers in space: does geography matter? J Ind Econ 64(2):295–335
75. Manova K, Yu Z (2016) How firms export: processing vs. ordinary trade with financial frictions. J Int Econ 100(C):120–137
76. Marin-Odio A (2014) Global value chains: a case study on Costa Rica, international trade. Technical Centre (ITC) Paper, Geneva, Switzerland
77. Mattoo A, Mulabdic A, Ruta M (2017) Trade creation and trade diversion in deep agreements. Policy research working paper IMF. Washington D.C.
78. Mazumdar K (2000) Causal flow between human well-being and per capita real gross domestic product. Soc Indic Res 50(3):297–313
79. Melitz MJ (2003) The impact of trade on intra-industry reallocations and aggregate industry productivity. Econometrica 71(6):1695–1725
80. Morris M, Staritz C (2014) Industrialization trajectories in madagascar's export apparel industry: ownership, embeddedness, markets, and upgrading. World Dev 56:243–257
81. Morris M, Staritz C (2017) Industrial upgrading and development in Lesotho's apparel industry: global value chains, foreign direct investment, and market diversification. Oxf Dev Stud 45(3):303–320
82. Muco M (1997) Economic transition in Albania: political constrains and mentality barriers, individual fellowship program, NATO. Faculty of Economics & Business, Univerisity of Tirana, Albania
83. Nordas HK (2004) The global textile and clothing industry post the agreement on textiles and clothing. World Trade Organization Publications, Geneva

84. Nunn N, Trefler D (2013) Incomplete contracts and the boundaries of the multinational firm. J Econ Behav Organ 94(1):330–344
85. Nuruzzaman AH, Rafiq A (2010) Is Bangladeshi RMG sector fit in the global apparel business? Analysis of the supply chain management. South East Asian J Manag 4(1):50–57
86. Ohno K (2010) Avoiding the middle income trap: renovating industrial policy formulation in Vietnam. ASEAN Econ Bull 26(1):25–43
87. Pathikonda V, Farole T (2017) The capabilities driving participation in global value chains. J Int Commer Econ Policy 8(1):1–26
88. Pierola MD, Fernandes AM, Farole T (2018) The role of imports for exporter performance in Peru. World Econ 41(2):550–572
89. Pietrobelli C, Rabellotti R (2011) Global value chains meet innovation systems: are there learning opportunities for developing countries. World Bank, Washington D.C.
90. Rahman M, Hamid M, Khan M (2015) Determinants of bank profitability: empirical evidence from Bangladesh. Int J Bus Manag 10(8):135–149
91. Rodrik D (2018) New technologies, global value chains, and developing economies. In: Pathways for prosperity commission background paper series, No. 1. Oxford, United Kingdom
92. Romano P, Vinelli A (2004) Quality management in a supply chain perspective, strategies and operative choices in a textile apparel network. Int J Oper Prod Manag 21:446–460
93. Rothenberg AD, Bazzi S, Nataraj S, Chari AV (2017) When regional policies fail: an evaluation of Indonesia's integrated economic development zones. Working paper, RAND Corporation, Santa Monica, CA.
94. Škabić IK (2019) The drivers of global value chain (GVC) participation in EU member states. Econ Res Eonomska Istraživanja 32:1204–1218. https://doi.org/10.1080/1331677x.2019.1629978
95. Taglioni D, Winkler D (2016) Making global value chain work for development. The World Bank, Washington D.C.
96. Todeva E, Rakhmatullin R (2016) Industry global value chains, connectivity and regional smart specialisation in Europe. An overview of theoretical approaches and mapping methodologies. JRC Science Policy Report, European Union, EUR 28086 EN. https://doi.org/10.2791/176781
97. UNCTAD (2017) World investment report 2017 investment and the digital economy. United Nations Publication
98. UNCTAD (2013) World investment report 2013: global value chains: investment and trade for development. United Nations Publication
99. United Nations Industrial Development Organization (UNIDO) (2018) Global value chains and industrial development, lessons from China. United Nations Publication, South-East and South Asia
100. World Bank (2019c) Rising uncertainties western balkans regular economic report No. 16, Washington D.C.
101. World Bank (2019b) World development report 2019: the changing nature of work. Washington, D.C.
102. World Bank (2019a) Global value chain development report. Technological innovation, supply chain trade and workers in a globalized world. World Bank, Washington, D.C.
103. World Bank (2017b) In Bangladesh, empowering and employing women in the garments sector. World Bank, Washington, DC. Retrieved from: https://www.worldbank.org/en/news/feature/2017/02/07/in-bangladesh-empowering-and-employing-women-in-the-garments-sector
104. World Bank (2017a) Global value chain development report. Measuring and analyzing the impact of GVCs on economic development. World Bank, Washington, D.C.
105. World Bank (2017) Measuring and analyzing the impact of GVCS on economic development. World Bank Publication, Washington DC
106. World Bank (2013) Western Balkans regional R&D Strategy for innovation under the Western Balkans Regional R&D strategy for innovation. World Bank Technical Assistance Project, under financing of European Commission DG ENLARG-TF011064

107. Xing Y, Detert N (2010) How the iPhone widens the United States trade deficit with the People's Republic of China. ADBI working paper series No. 257
108. Yameogo ND, Jammeh K (2019) Determinants of participation in manufacturing GVCs in Africa: the role of skills, human capital endowment, and migration. Policy research working paper 8938. World Bank, Washington, D.C.
109. Yin RK (2003) Case study methodology, 3th edn, Sage Publications Inc., New York

Chapter 4
Integration of Emerging Economies in the Global Value Chains of the Clothing Industry

Abstract The clothing industry is widely considered for developing countries as the first step of industrialization and integration into global value chains. Given its importance, this chapter presents the main features of the clothing industry including recent developments on its global value chain and the production cycle oriented towards the fast fashion. With regard to developing countries, this chapter presents the data centric approach to evaluate the integration of emerging economies into global value chains of the clothing industry. Time series analysis on data sets from EORA, ICIO, and World Integrated Solutions are examined to identify the key determinants to facilitate a higher degree of participation in international trade. A special focus in this chapter is given to the Western Balkan region.

Keywords Clothing industry · Location advantages · Cost advantages · Fast fashion · Agile value chain

4.1 Introduction

The clothing industry is globally spread from one continent to the other. It was one of the most protected of all industries, ranging from agricultural subsidies on input materials (cotton, wool, rayon) to a long history of quotas under the Multi Fibre Arrangement (MFA) institutionalized in 1974, which ruled international trade in the clothing industry for two decades and Agreement on Textiles and Clothing (ATC) in 1995. The MFA/ATC restricted exports to major consuming markets by imposing country limits (quotas) on the volume of certain imported products [36, 46]. Trade restrictions have contributed to the international fragmentation of GVC in the clothing industry, in which developing countries having a cheap labor force engage mostly on assembly activities of imported inputs and re-export of final products. The removal of quotas in 2005 marked the end of over 30 years of restricted access to the European Union and North America markets. This caused a tremendous flux in the global geography of production and trade together with a restructuring of strategies to adjust production in line with the new economic and political global environment [64, 75].

© Springer Nature Switzerland AG 2020
J. Kacani, *A Data-Centric Approach to Breaking the FDI Trap Through Integration in Global Value Chains*, Lecture Notes on Data Engineering and Communications Technologies 50, https://doi.org/10.1007/978-3-030-43189-1_4

The global clothing industry is a multi-billion industry and it is considered a source of economic development and industrialization. The clothing industry is also a forerunner of globalization, as it was one of the first industries to adopt a global dimension, incorporating developing countries into the global value chain. Today, the industry is undergoing a profound change reflecting the transformation of the global economy [76, 77]. The clothing industry has closely mirrored the general trends of consumption and taste in the last decade and has become more customer oriented by putting special emphasis on the design, branding, and marketing functions in the GVC of the clothing industry. Lead firms in developed countries are mostly branded manufacturers that dominate the GVC in the clothing industry undertake and perform these functions. Developing countries typically participate through provision of services mostly on labor-intensive production processes. Participation in the GVC of the clothing industry is a way for developing countries to: (i) attract investment, (ii) increase technological capability, (iii) build industrial capacity, (iv) foster economic growth, (v) generate employment, and (vi) develop a network of linkages. These benefits make the clothing industry a key player in the global environment [58].

Clothing manufacturing enterprises face a variety of challenges. Since the removal of quotas, the global clothing industry is faced with overcapacity that is creating intense competition from low-cost countries [21]. Quotas created too many factories in too many locations, and now these factories are competing for orders. In the short term, this has significantly increased the standards as customers are asking for quality products, more services, and faster turnaround times, all for lower costs. Suppliers must meet demand to keep orders, increase volume, and reduce costs while facing an ongoing consolidation in the retail sector giving more power to global customers including retailers, global brands, and large manufacturers that outsource their production [55]. Customers are aware and sensitive not only to clothing manufacturing but also for utilizing questionable labor practices to reduce costs. Monitoring of these labor conditions is very difficult especially for home workers and for those in the informal market. However, in the clothing industry the importance of skills has increased particularly for supervisors, technical positions, and managers [65, 81]. Increasing requirements from customers have raised the demand for relatively skilled workers and high labor productivity. Additional challenges in the clothing industry include constant fluctuations in commodity prices, seasonal changes and trends, and with speed becoming essential in bringing a product from concept to shelf as quickly as possible. In addition, the role of outsourced manufacturing has created value chains spanning in countries and continents with inherent challenges of management and oversight [38].

This chapter begins with a detailed description of the textile-clothing global value chain and subsidiary upgrading followed by a presentation of the clothing industry in Albania. In addition, through OLS model estimates this chapter presents the integration of the clothing industry in Albania into the GVCs in the same industry. The chapter continues with the presentation of the Façon Package and concludes with the advantages offered by Albania to foreign clothing manufacturers.

4.2 The Textile-Clothing Value Chain

According to Anner [5] and Bair and Gereffi [8], the current internalization of markets led to the establishment of GVCs. Moore specifically the global value chain in the clothing industry consists of links that show interrelated activities such as the production and distribution of goods and services (see Fig. 4.1). The GVC in the clothing industry extends from raw materials (e.g., cotton, or petrochemicals), to production of natural or synthetic fibres and textiles, then to the design, cutting, assembly, laundering of final goods, and, finally to the distribution, marketing, and retailing of final goods to end customers. This chain refers to the production and distribution processes across borders through close cooperation of separate firms, outsourcing and off-shoring activities to developing countries [33].

According to the International Labour Organization (ILO) [44] and Chan [18], the GVC in the clothing industry encompasses customers, retail stores, distributions centers, manufacturing plants, textile plants, and raw materials. The dotted lines represent the flow of information, while the solid lines represent the flow of goods. The information flow starts with the customer and forms the basis of what is being produced and when. It is also worth noticing that information flows directly from retailers to textile plants in many cases [34]. The textile sector produces for clothing manufacturing and for household use. In the first case, there is direct communication between retailers and textile mills when decisions are made on patterns, colors and material. In the second case, textile mills often deliver household appliances directly to retailers [2]. The textile-clothing GVC has four main segments (see Fig. 4.1).

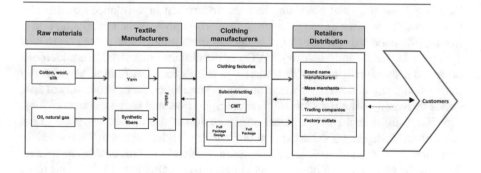

Fig. 4.1 Textile-clothing global value chain. [–] flow of goods and [...] flow of information. *Source* Author's adaption based on Fernandez-Stark et al. [28] and Gereffi and Memedovic [35]

- **Raw materials** refer to materials can be in the form of natural and synthetic fibers. Natural fibers like cotton or wool come from agriculture inputs while synthetic fibers are generated from other natural resources like oil and gas [37, 60].
- **Textile manufacturers** include firms that specialize in production of various fabrics by spinning, weaving, dyeing, and printing raw materials. Production is characterized by large scale and sophisticated machines. Textile manufacturers are usually large firms able to make high level of investments required to purchase and maintain sophisticated machines used for production of fabric. This is the most capital-intensive segment of the value chain [22, 74].
- **Clothing manufacturers** include firms that assembly the fabric obtained from textile manufacturers. Compared to textile manufacturing, production of cloths is based on less sophisticated machines (sewing and cutting). Clothing manufacturers are usually of all sizes ranging from large firms employing thousands of workers to small subcontractors. This segment of the value chain is the most labor intensive.
- **Retailers/Distribution** include firms that are: (i) brand manufacturers (Levis and Tommy Hilfiger), (ii) brand marketer (Fruit of the Loom and Zara), (iii) mass merchants (Walmart, Tesco, C&A), (iv) specialty stores (Gap, Benetton, H&M), (v) trading companies that work with catalogue delivery (Otto Group), (vi) factory outlet (TK&Max). These firms are regarded as "lead firms" of the value chain as they dominate the export and marketing of goods produced by clothing manufacturers. These firms are the ones that reach the end customer [30, 36].

To continue, the textile-clothing GVC is a typical "buyer-driven chain", where lead firms coordinate global production of clothing in relation to final customers on the one hand, and local industries in developing countries on the other hand. Most value in the clothing industry is associated with the control of key functions such as branding, designing, and marketing that are undertaken in developed countries by lead firms. These firms obtain the highest profits and possess the negotiating power from the demand side [32, 36, 39]. Changes in trade policy have given rise to new dynamics in the clothing industry. GVCs in the clothing industry have undergone profound restructuring to meet new market demands for "fast fashion", marked by rapid shipments characterized by high quality requirements, the right quantity of final goods delivered, at the right location, on the right time, and in the right conditions. This has resulted in the geographical fragmentation of the various segments of the value chain. Today, the GVC in the clothing industry is oriented towards waste reduction, time compression, and flexible response, reduction in unit costs emphasizing this way the importance of both intra and inter firm coordination [15, 54]. This trend recognizes that the way to competitive advantage lies in the GVC of the clothing industry [42].

Lead firms are the main players in the GVC of clothing industry. They respond much faster to retailer demands transforming the time involved in meeting orders into a cost. Currently, only few retailers continue to manufacture in the industry as most are focused in outsourcing and subcontracting arrangements [24, 43].

The emergence of GVCs has major policy implications for economic growth in developing countries. For many industries, the spread of clothing manufacturing across various countries has lowered the costs of production of final goods and has increased the productivity of labor and capital. According to Taglioni and Winkler [73] and Baldwin [9], the spread of GVCs has two main consequences for developing countries. Firstly, it has created a path through which countries can industrialize at a much earlier stage of development as lead firms choose to outsource fractions of the value chain to countries where labor is cheaper or where locational advantages allow a competitive cost advantage on the whole value chain [63, 76]. Secondly, it allows suppliers to meet standards and regulations that permit them to access markets in developed countries, and to utilize advanced technology that would not otherwise be available [14, 75].

An example of how the GVC in the clothing industry operates is that of US retailers. They typically replenish their stores on a weekly basis. Point of sales data are extracted and analyzed over the weekend and replenishment orders are placed with the manufacturer on Monday morning. The manufacturer is required to fill the order within a week, which implies that he/she will always have to carry larger inventories of finished goods than the retailer. How much larger depends on his own lead-time and volatility in demand [19, 41]. Frequent fluctuations in demand and the greater the variety of products in style, size and color the larger the inventory has to be. On the other hand, a shorter manufacturing time leads to better the demand forecasts. Upon receiving the replenishment order, the manufacturer will fill it from its inventory and then based on the gap between remaining inventory and the desired inventory level, will place orders to production plants, located in different territories [51, 66]. Retailers may order large quantities of a given product from a number of producers located in several low-wage countries. In order to ensure that goods are similar and can sell under the same label, retailers often buy fabric and accessories in large quantities to provide the same inputs to all clothing manufacturers. In addition, retailers also specify the design, style and sizes to suppliers and assist manufactures during production in order to obtain the desired quality [21, 27].

To continue, the GVC in the clothing industry needs to be both lean and agile, giving rise to the "leagile" chain. Agility can be regarded as the ability of an organization to respond rapidly to changes in demand both in terms of quantity and in terms of variety [15, 40]. Agility is an operational capability that encompasses organizational structures, information systems, logistics processes, and particular mindsets. A key characteristic of an agile organization is flexibility. In GVCs, agility implies the ability to react quickly to changes in market demand [3, 7].

For Taglioni and Winkler [73] and Martin [55], an agile GVC has a number of distinguishing characteristics (see Fig. 4.2). Firstly, the agile value chain has to be market sensitive. In this case, the GVC is able to understand and respond to real demand and not only to the inventory forecasts based on past sales and shipments. An agile GVC obtains data on demand directly from the point of sale permitting players of the value chain to identify the needs of the market and to respond directly to them.

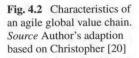

Fig. 4.2 Characteristics of
an agile global value chain.
Source Author's adaption
based on Christopher [20]

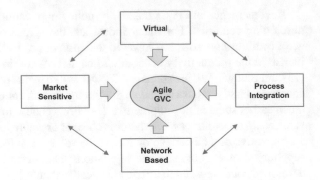

Secondly, the use of information technology to share data between buyers and suppliers has given a virtual trait to GVCs. Virtual GVCs are information based rather than inventory based. The global online network has enabled organizations in GVCs to act upon the same data to generate the real demand rather than be dependent on the history of orders [2].

Thirdly, process integration is another characteristic GVCs must possess in order to be considered agile. Process integration means a strong cooperation between buyers and suppliers including joint product development, common production systems, and shared information [57]. Along with process integration comes joint strategy determination, strong buyer-supplier teams, transparency of information, and even open-book accounting. This form of co-operation has a high prevalence in value chains as companies focus on managing their core competencies and outsource all other activities [25, 51]. In volatile markets, alliance between suppliers and buyers is both inevitable and essential for maintaining competitiveness.

The forth characteristic of an agile GVC regards association of firms participating in the GVC linked together as a network. The network allows to gain a larger market share as organizations part of a network are better structured and are able to manage more active relationships with end customers. It can be argued that in today's unpredictable global markets sustainable advantage lies in being able to intensify the strengths and competencies of partners in the network aiming to achieve greater responsiveness to market needs [10, 68].

In addition to agility, GVCs need to be lean. A lean GVC works best in high volume, stable market environment, and predictable demand. The focus in the management of GVCs management is elimination of all waste, including time, to enable the creation of a stable order schedule. Lean retailers require rapid replenishment of products, and shipments that restrict requirements in terms of delivery times, order completeness and accuracy. Key to this is the use of bar codes, EDI and shipment marking [78]. However, lean strategies are effective in markets where demand is stable and products are standard. Where volatility dominates markets then strategies will inevitably become agile [16].

A dominant market strategy is based on the concept of "leagile" GVC [20, 56, 58]. "Leagile" takes the view that a combination of lean and agile approaches are essential to compete in a challenging business environment. Agility and leanness are not opposing approaches rather they work best if combined. The issue is not to choose between agility and leanness, rather it relies in selecting and efficiently integrating appropriate features of each strategy into the particular value chain [17, 65].

4.3   The Main Features of the Clothing Industry

The clothing industry is one of the most geographically fragmented industries. The geographical dispersion has proved beneficial for newly developed and industrialized countries that used clothing industry as the first step for their economic development. Low barriers to entry in production characterize this industry. However, referring to Rodrik [67] it is easy to enter production in the clothing industry due to low capital intensity but it is difficult to generate high profits segments as they are controlled by lead firms. The low barriers to entry give this industry its transitory nature and turn it into a relatively mobile industry [47].

A key characteristic of the clothing industry is its orientation towards labor-intensive processes, which generate entry-level jobs for unskilled labor in developed and developing countries. Labor costs are the main production cost in the clothing industry (see Fig. 4.3). The major advantage of low labor costs lies in the manufacturing of basic items that sell largely because of price difference from fashion garments

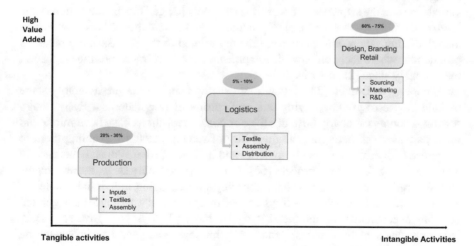

Fig. 4.3   Average cost structure in the clothing industry according to value adding activities. *Source* Author's drawing based on UNIDO [77] and from interviews in the case studies. Reference is made to the cost structure obtained during interviews with high management in Naber Konfeksion sh.p.k and Shqiperia Trikot sh.p.k

in which design and style are important [24, 30]. Production that corresponds to the lowest value added activities in the GVC of the clothing industry accounts for 20–30% of total costs. Given their low value added these activities are located in developing countries taking advantage of the low production costs. The high share in the cost of overall production is associated with high value added activities remains in developed countries under the supervision of lead firms [66]. Unlike producer-driven chains, where profits come from scale, volume and technological advances, in the buyer-driven textile-clothing value chain, profits are generated from combining high value research, design, sales, marketing, and financial services that allow lead firms to act as an intermediary between producers and customers [5, 35, 43]. The main firms[1] are powerful players in the textile-clothing value chain as they control high value adding activities distributed along the value chain. The most notable feature of value adding activities regards the greatest gains generated in the services occurring before and after production [19, 50].

Employment in the sector has been particularly strong for women in poor countries, who previously had no income opportunities other than the household or the informal sector. Around 80% of employees in the clothing industry are female that are mainly unskilled or semi-skilled [28, 74]. Employment in the clothing industry tends to be also of a temporary nature as it fluctuates depending on the variations of the market demand. Despite that sometimes employees face working conditions falling outside national regulations clothing manufacturing firms do not experience a significant shortage of workers.

The basic production knowledge in the clothing industry has not changed much over past decades. Technology used in clothing manufacturing leaves little room for change as this industry depends on complex manual operations. Even though basic knowledge and the sequence of operations have not changed much, innovations have improved efficiency at each stage of production and have enhanced coordination in manufacturing. These innovations are mainly related to the pre-assembly phase of production, where technological developments have been more prominent than at the assembly stage [26, 60, 69].

According to Dicken [24], know-how developments in the pre-assembly stage include processes grading, laying out, and cutting of raw materials. Even though sewing accounts for nearly 80% of all labor costs innovations have been minor and mostly focused on increasing the flexibility of sewing machines enabling them to recognize oddly shaped pieces of material and indicating where adjustments need to be made during the sewing process [63]. In the post assembly stage, advancements include production delivery systems focusing on conveying individual articles to employees through a conveyer belt system and more efficient warehouse management resulting in cost reduction and time saving. Moreover, Phu [64] and Bruce and Daly [16] argue that in clothing industry modern technology is adopted even in developing and poor countries at relatively low investment costs.

---

[1]Main firms are primary sources of material inputs, technology transfer, and knowledge in the organizational networks of the clothing industry. They use different networks and sources in multiple regions of the world.

4.3.1 Production Modalities in the Clothing Industry

With reference to ILO [44] and Goto et al. [39], there are four main production stages in the clothing industry (see Fig. 4.4).

4.3.1.1 Cut Make Trim (CMT)

It is the most basic production stage in the clothing industry, in which production plants are provided with imported inputs for assembly. The clothing manufacturer is responsible for cutting, sewing, supplying trim, and/or shipping the ready-made garment. The buyer purchases the fabric and supplies it to the manufacturer, along with detailed technical specification required for production. The manufacturer works with a number of different customers and their operations consist of an order-by-order basis [8, 40].

4.3.1.2 Full Package

It is a stage in which the manufacturer takes responsibility for all production activities, including the CMT activities, as well as finishing and distribution. The firm must have logistics capabilities, including procuring and financing, and the required raw materials needed for production. In some cases, the customer specifies a set of textile firms from which the manufacturer must purchase materials, while in other cases, the firm is responsible for establishing its own network of suppliers [37, 52]. The manufacturer is also often responsible for downstream logistics, including packaging for delivery to the retail outlet and shipping the final product to the customer at an agreed selling price [35]. The customer typically provides to the manufacturer product specifications and designs, but the customer is not involved with the details of the manufacturing process, such as pattern making. Full package service providers can range from operations in single production to global suppliers, which have multiple production centers and work on multiple product ranges [17, 50].

4.3.1.3 Full Package with Design

The manufacturer is responsible for full package services and for the design. A manufacturer that provides full package with design carries out all steps involved

Fig. 4.4 The four main stages of production in the clothing industry. *Source* Author's drawing

in the production of an article including design, fabric purchasing, cutting, sewing, trimming, packaging, and distribution [31]. Typically, the manufacturer organizes and coordinates the design of the product, approves the samples, purchases raw materials, completes production and, in some cases, delivery of finished products to end customer. Full package with provision of design is common for private-label retail brands [53, 77].

4.3.1.4   Original Brand Manufacturing (OBM)

It is the most advanced stage of production in the clothing industry as the manufacturer realizes branding of products including promotion and sale of its own branded products. The main products in the clothing industry are prepared in accordance with one of the main characteristics of the clothing industry that of the "fast fashion".

To continue, the market in the clothing industry is divided into differentiated and standardized products. Differentiated products refer to high quality and fashionable products. Firms manufacturing differentiated products are characterized by technology that is more advanced and a high degree of flexibility [70].

The competitive advantage of firms in this market segment is related to the ability to produce designs that capture and influence the taste and preferences of customers. On the other hand, firms manufacturing standardized products such as t-shirts, uniforms, underwear are less technologically advanced and are mostly located in developing countries, often in export processing zones and operating under the outward processing regime [19, 72].

According to Bamber et al. [10], Dicken [24] and Daspal [23], depending on price[2] and income[3] elasticity of customer demand, products in the clothing industry are broadly divided into three types (see Fig. 4.5):

- **Basic** are products that usually have a low volatility of demand with an income and price elasticity lower than one.
- **Fashion basic** are products that have a moderate volatility of demand with an income and price elasticity greater than one.
- **Fashion** are products that exhibit a high volatility of demand with an income and price elasticity greater than one which is higher than in the case of products falling into fashion-basic category. These products are highly priced beyond products that fall into necessities. The level of necessities demand for clothing products increases less rapidly than the growth of income.

---

[2]Price elasticity of demand reflects the changes in the quantity demanded by customers when the price of the product changes. A price elasticity <1 means that the demand is inelastic and that the quantity demanded would not be much affected by changes in price. A price elasticity >1 means that changes in price will significantly affect the quantity demanded of the product (McConnell et al. 2011).

[3]Income elasticity of demand reflects the changes in the quantity demanded by customers when their income fluctuates. An income elasticity <1 indicates that the quantity demanded will not be much affected. An income elasticity >1 shows that fluctuations in income will affect the quantity demanded (McConnell et al. 2011).

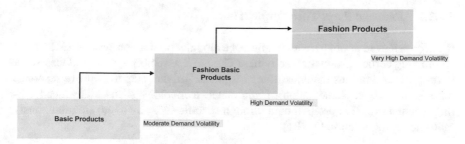

Fig. 4.5 Products in the clothing industry with reference to volatility of demand. *Source* Author's drawing

4.3.2 Towards Fast Fashion

The idea of fast fashion started in the United States in the 1970s when retailers such as Wal-Mart constrained suppliers to implement information technologies for exchange of data on sales, to adopt standards for product labelling and methods for material handling. This ensured quick replenishment of clothing, which in turn allowed the retailer to offer a broad variety of clothing without holding a large inventory [1, 58]. The success of this approach in the clothing industry spread to other industries shifting the competitive advantage of manufacturers from being simply an issue of production costs to becoming an issue of costs in combination with lead time and flexibility [50, 54].

In meeting frequent orders of lead firms and in satisfying customer preference manufactures are constrained to deliver final products in the shortest possible time. Therefore, they have implemented the quick response system, giving rise to fast fashion. Many clothing manufacturers have introduced changes in internal processes and have oriented them towards quick delivery. The quick response system (QRS) permits manufacturers to deliver finished articles in the shortest time possible. QRS emphasis upon flexibility and product velocity in order to meet the changing requirements of a highly competitive, volatile, and dynamic marketplace. QRS encompasses a strategy, structure, culture and set of operational procedures aimed at rapid information transfer resulting in profitable production activities. Implementing a QRS requires a degree of organization within the firm and a good coordination with the customer. Referring to ILO [44], clothing manufacturers need to address simultaneously four key elements so to ensure that delivery times are at a minimum: (i) reduction of plant throughput time, (ii) implementation of electronic data interchange, (iii) improvements in inventory control systems, and (iv) ensuring minimum transport times.

4.3.2.1   Reduced Plant Throughput Time

It can be achieved through a combination of a modular production systems (MPS) and automating certain production activities. The MPS assembly system reduces work in progress and fastens manufacturer's response. Another step to improve response time is through implementation of automated technology such as advanced cutting machines that are programmed through specific software to cut simultaneously multiple layers of fabric [51, 81].

4.3.2.2   Implementation of Electronic Data Interchange

It is another important method applied to streamline the processing of payments and contractual agreements between the manufacturer and retailer through implementation of electronic data interchange (EDI). This system can reduce the time required to perform processes like orders, payments, and contracts to hours and even minutes. Automating the process reduces costs in human resources and results in fewer errors, often due to human mistakes [27, 59]. This is especially important for retailers and suppliers of different nationalities, where language is often a barrier and orders can be easily misinterpreted.

4.3.2.3   Improvement of Inventory Control Systems

This is the third method of reducing lead times and maintaining an organized and efficient inventory control system. One problem faced by clothing manufacturers is excessive amounts of inventory arising from the varieties in the same product including design, style, and colors. Currently, more clothing manufacturers are incorporating point of sale (POS) data from retailers to determine their optimal inventory levels [23, 78].

4.3.2.4   Reduced Transport Time

It refers to many transport delays beyond the control of clothing manufacturers like poor roads, bad weather and delays at customs, they need to take additional actions to ensure that shipments to distribution centres of retailers arrive in the minimum amount of time possible [14, 76]. A common way is to subcontract a transporting company. Even though there are additional costs involved, subcontractors can ensure quick fulfilment of orders making the extra investment worthwhile [24, 46]. Another action is to ensure that shipping containers and cartons are labelled according to agreed standards helping retailers to handle large quantities especially in large distribution centres where orders from hundreds of different suppliers can arrive at the same time.

4.3.3 Types of Upgrading in the Clothing Industry

Upgrading in the clothing industry refers to the ability to perform activities that are more complex by improving the efficiency of production processes and by realizing progressively more complex product lines. Upgrading allows clothing manufacturers to move into higher segments of the value chain. For clothing manufactures, it means to move from assembly to original brand manufacturing. What follows in this section are the three main types of upgrading together with the main factors that affect upgrading in clothing manufacturers.

Referring to Grumiller et al. [40], Morris and Staritz [58], Kaplinsky and Morris [49] and Humphrey and Schmitz [41], upgrading can be achieved by:

- improving the efficiency of production processes (process upgrading)
- adding new product lines due to improvements in design or technical specifications (product upgrading)
- taking on new functions which require higher level of skills, knowledge, and intensity (functional upgrading).

4.3.3.1 Process Upgrading

It refers to application of new technology or rearranging existing production systems. Nevertheless, several empirical studies have indicated that transfer of advanced technologies through linkages with production and distribution networks coordinated by international buyers have become important in process upgrading [30, 39].

4.3.3.2 Product Upgrading

It refers to a shift into product lines, which are normally more difficult to produce because of differences in technical specification and input materials. For instance, a subsidiary may experience product upgrading by shifting from production of casual woven shirts to expensive suits. The ability of clothing manufacturers to generate higher value products is strongly correlated to the extent manufactures are able to achieve upgrading in production processes [8, 63].

4.3.3.3 Functional Upgrading

It refers to moving into more complex functions in a particular value chain. In essence, functional upgrading has to do with shifting towards more knowledge and skill-intensive functions in the value chain [38, 58]. In the clothing industry, such functions include product design, material sourcing, branding, and marketing. Moving up the chain into higher value-added functions entails organizational changes in distribution and production, which is probably most difficult to achieve [43, 60, 77]. For example, the CMT modality consists in functions that are mostly dependent on unskilled or semi-skilled labor and, therefore, is one with the lowest value-added contents. Suppliers can functionally upgrade and shift to OBM, by integrating higher knowledge-intensive functions such as sourcing, designing, branding and marketing. Functional upgrading depends heavily on the capacity of manufacturers to handle the increasingly complex functions and to a certain extent on the willingness to delegate such functions to manufacturers.

Nevertheless, when clothing manufacturers in developing countries upgrade in terms of processes and products, this does not mean that those manufacturers are moving up along the value chain and entering into higher value-added activities. This type of upgrading occurs within the same operations such as higher efficiency levels within the CMT assembly (process upgrading) or lead to production of more sophisticated products within the same product category (product upgrading). However, this can be achieved through:

4.3.3.4 Marketing and Networking

According to Anner [5] and Figueiredo [29], clothing manufacturers need to put considerable efforts oriented toward marketing and networking to form alliances with other firms and international organizations dedicated to development, research, and best practices. Membership in such networks is encouraged through participation in international trade shows to increase visibility to potential buyers.

4.3.3.5 Investments in Technology

Investments are needed to upgrade machines used in production as well as logistics and information technologies that enable manufacturers to become more integrated into the networks of the global value chain of the industry [31, 44, 74].

4.4 The Clothing Industry in Albania

In this section, a comparison of Albania with neighboring countries is followed by the situation of the clothing industry in the country. In addition, the integration of Albania in the GVCs of the clothing industry is estimated based on time series data and OLS models. This section is concluded with the industrial policy in the clothing industry with special focus on the Façon Package accompanied with location advantages available to foreign investors in Albania.

4.4.1 The Business Environment in Albania Compared to Countries in the Western Balkans

This section presents an overview of the doing business indicators in Albania and the remaining countries in the Western Balkan. The doing business indicators in a host territory are among the variables that foreign investors look into before deciding to locate production. The data is obtained from the World Bank Doing Business Report (2016–2020) by selecting the indicators that are of relevance to the doing business in the clothing industry.

With regard to the ease of doing business, Albania lags behind other regional countries like FYR Macedonia and Montenegro. In 2019, Albania lost five places compared to 2017 mostly attributed to the change in the fiscal regime from a flat system into a progressive one and the political instability experienced within the last two years (see Table 4.1). Albania is well placed for protecting of foreign investors not only in the region but also worldwide with the highest score achieved in 2017 when it globally ranked in the 19th place. However, even in this area the country has experienced a decline of seven places in 2019, mostly from the turbulent political occurrences in Albania and the inability to open negotiations for membership with European Commission.

For clothing manufacturing enterprises, trading across borders is a key indicator to determine the costs encountered when shipping final goods (see Table 4.2[4]). In 2020 Albania, is doing relatively well on trade across borders with regard to export costs. In 2020 even though in monetary terms Albania had the lowest costs, it ranked poorly with regard to time required for documentary and border compliance procedures. These figures are a threat to the favorable geographical position of Albania given that among countries in WB it has the shortest distance to export into the European market. The high number of days reflects the lengthy custom procedures, on of the main obstacles for exporting of manufactured goods identified in this research.

Table 4.1 Doing business and protecting investors in Western Balkans countries (2017–2019)

Year	2017		2018		2019	
Indicator (Rank)	Doing business	Protecting investors	Doing business	Protecting investors	Doing business	Protecting investors
Albania	58	19	65	20	63	26
North Macedonia	10	13	11	4	10	7
Serbia	47	70	43	76	48	83
Bosnia Herzegovina	81	81	86	62	89	72
Montenegro	51	42	42	51	50	57

Source World Bank Doing Business 2019

---

[4]The figures are based on the recent World Bank Doing Business Report (2020).

Table 4.2 Trade across borders export[a] in the Western Balkan countries (2019–2020)

Doing business trade across borders (DB methodology 2016–2020)										
Year	2020					2019				
Exports	Albania	Serbia	North Macedonia	Bosnia Herzegovina	Montenegro	Albania	Serbia	North Macedonia	Bosnia Herzegovina	Montenegro
Documentary compliance (h)	6	2	2	4	5	6	2	2	4	5
Border compliance (h)	9	4	9	5	8	9	4	9	5	8
Cost to export: documentary compliance (USD)	10	35	45	22	26	10	35	45	92	67
Border compliance (USD)	55	47	103	70	85	55	47	103	106	158

Source World Bank Doing Business 2020

[a]Doing Business measures the time and cost (excluding tariffs) associated with three sets of procedures—documentary compliance, border compliance and domestic transport—within the overall process of exporting or importing a shipment of goods. Time is measured in hours, and 1 day is 24 h. Costs are reported in U.S. dollars. Insurance cost and informal payments for which no receipt is issued are excluded from the costs recorded. Documentary compliance captures the time and cost associated with compliance with the documentary requirements of all government agencies of the origin economy, the destination economy and any transit economies

A similar scenario occurs also with the cost of import (see Table 4.3). In 2020, Albania had the lowest cost among regional countries in monetary value while the time required is longer than the remaining WB countries. This indicator is key to clothing manufacturers as the majority of raw materials are imported. In addition, the time required to import in Albania remains one of the longest in the region even more than Bosnia Herzegovina, that does not have the access to the EU market (through major ports) like Albania. This shows that Albania needs to simplify its customs procedures.

In addition, Albania had the lowest average wage among the regional countries (see Fig. 4.6). The average wage before tax in 2020 in Albania is 25% lower than in Bulgaria and 21% cheaper than North Macedonia. On the other hand, in Croatia the average wage before tax is 2.47 times higher than in Albania. The data reflects the availability of the cheap labor force in Albania, one of the main locational advantages identified during fieldwork in the four clothing manufacturing subsidiaries.

4.4.2 The Situation of the Clothing Industry in Albania

From early 1960s until 1989, the textile, clothing, and leather/footwear industries were among the key players of the manufacturing industry in the Albanian economy. Enterprises operating in this industry were state owned and operated in a centralized economy dominated by a socialist regime.

Production within the 30-year period was very diverse as the domestic economy was self-sustaining. Products ranged from raw material (fibers, fabrics, etc.) used in manufacturing to finished products (dresses, coats, upholstery, etc.). The clothing industry before the 1990s was self-sustained as the whole production cycle was undertaken in the country. Even though at basic level the clothing manufacturing included the design of cloths and distribution throughout the country.

With the fall of the socialist regime and the establishment of the market economy, state owned enterprises were privatized and manufacturing enterprises including those in the clothing industry started to produce through the inward processing regime.[5] This regime is also called Façon manufacturing [11]. After the 1990s even though the clothing industry is not self-sustained and it depends on the import of intermediary products, it is gradually spanning to more value added services like R&D and provision of services (Fig. 4.7).

According to the World Bank Enterprise Investment Survey (2007), since the start of their operational activity in Albania in the early 1990s foreign owned clothing manufacturing enterprises have enriched the variety of articles they manufacture.

---

[5]Inward processing regime allows imported raw materials or semi-manufactured goods to be processed for re-export without the requirement that the manufacturers have to pay customs duty and VAT on the goods being used.

http://ec.europa.eu/taxation_customs/customs/procedural_aspects/imports/inward_processing/index_en.html.

Table 4.3 Imports in the Western Balkan countries (2019–2020)

Doing business trade across borders (DB methodology 2016–2020)

Year	2020					2019				
Imports	Albania	Serbia	North Macedonia	Bosnia Herzegovina	Montenegro	Albania	Serbia	North Macedonia	Bosnia Herzegovina	Montenegro
Documentary compliance (h)	8	3	3	8	6	8	3	3	8	10
Border compliance (h)	10	5	8	6	23	10	4	8	6	23
Cost to export: documentary compliance (USD)	10	35	50	27	60	10	35	50	97	100
Border compliance (USD)	77	52	150	109	306	77	52	150	109	306

Source World Bank Doing Business 2020

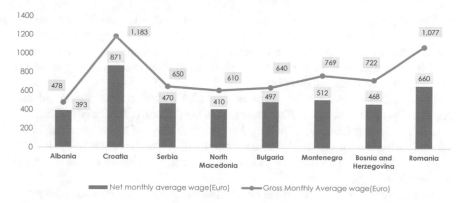

Fig. 4.6 Average monthly wage in Western Balkan countries and the recent joiners of the European Union 2019. *Source* National Statistical Institutes of each country

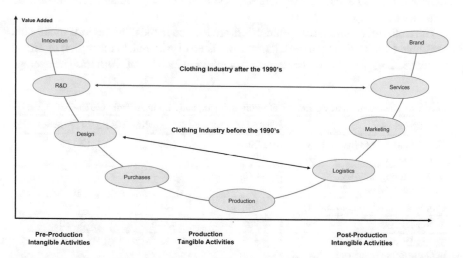

Fig. 4.7 The value adding activities in the clothing industry in Albania before and after the 1990s. *Source* Author's drawing based on Frederick and Daly [30]

Between 76.6% of foreign-owned clothing manufacturing enterprises reported to have introduced new products compared to only 37.5% of locally owned firms.

With reference to Kacani and Van Wunnik [48] on Italian owned clothing manufacturing enterprises, mentioned the proximity and the possibility to transport quickly finished articles in the EU market as key factors on deciding to move production in Albania. With reference to calculations of OECD [61, 62] the costs enterprises incur in shipping products from Albania to Italy is equal to 1.43% of the value of imports. The data is in line with the average shipping costs of 1.48% in the Western Balkans but it is substantially lower than 11.8% required to ship products from China.

Clothing manufacturers in Albania (local and foreign) incur higher production costs due to unreliable supply of electricity. They face many difficulties and are sceptical to use in production modern equipment or technologically advanced machines because they are afraid that unpredictable interruptions in power supply can permanently damage them [79].

Between 2015 and 2017, the share of the value added of the clothing industry generated as a percentage of GDP generated in the Albanian economy ranges from 1.31 to 1.27% (see Table 4.4). On the other hand, the share of the value added of the manufacturing industry in the Albanian economy fluctuates between 5.67% in 2015 and 6.13% in 2017. The share of FDI stock in manufacturing of cloths fluctuates between 5 and 10%.

According to the Chamber of Façon of Albania, in 2015, the number of clothing manufacturing enterprises in Albania was 608[6] enterprises that employed 43,264 workers (see Table 4.5). Foreign owned clothing manufacturing enterprises are 19.9% out of which 54% have Italian ownership, 14% have Greek ownership, and 4.9% have German ownership.

The clothing manufacturing enterprises are located in the main regions of Albania (see Table 4.6). Referring to the Chamber of Façon of Albania, in the region of Tirana are located 23.5% of the total enterprises employing 29.61% of workers. The average

Table 4.4 The manufacturing and the clothing industry, and FDI in Albania (2015–2017)

Description	2015 (Mln/EUR)	2016 (Mln/EUR)	2017 (Mln/EUR)
Gross domestic product (GDP) at market prices	11,206	11,504	12,119
Manufacturing industry contribution (value added)	636	654	744
Clothing industry (value added)	146	147	n/a
Total FDI stock	3,902	4,486	5,930
Total FDI stock in the manufacturing industry	574	601	591
Total FDI stock in manufacturing of cloths	5–10%	5–10%	5–10%
FDI inflows	851	1,001	1,043
FDI inflows in the manufacturing industry	6	11	15

Source National Accounts, INSTAT, Bank of Albania

---

[6]The number of enterprises is according to an initiative undertaken by the Chamber of Façon in Albania entitled "Creation of an integrated database for the Clothing Industry in Albania". The aim of the initiative was to identify the current active clothing manufacturing enterprises in Albania. The need for proper identification arose as the researcher and the project team identified that enterprises registered in the National Registration Center as clothing manufacturers were not engaged in production but only imported cloths sold in various shops around Albania.

Table 4.5 Enterprises, employment, and nationality in the Albanian clothing industry (2015)

Description	Total	Foreign	Italian ownership	Greek ownership	German ownership	Total employment	Total employment in foreign owned enterprises
Clothing manufacturing enterprises	608	121	66	17	6	43,264	18,248

Source The Chamber of Façon of Albania

Table 4.6 Regions of Albania with the majority of active clothing enterprises (2015)

No.	Description	Number of clothing manufacturing enterprises	Number of employees	Average number of employees
1	Tirana	143	12,811	90
2	Shkodra	34	3700	109
3	Durres	94	10,372	111
4	Berat	17	3673	223
5	Korca	46	3036	66

Source The Chamber of Façon of Albania

number of employees in the clothing manufacturing enterprises located in the region of Berat is 223, higher than in any other region.

In 2018, exports in the clothing industry have increased by 37% compared to 2015 (see Table 4.7). The increase of exports has caused an increase of 30% in the imports of cloths. Part of this increase is attributed to the import of intermediated goods used to produce final goods for exports that are exported mainly in the European Market.

Italy, Greece, Germany, and Spain are the main trade partners in the clothing industry. Italy remains the principal partner with 70.34% of the exports in the clothing industry in 2018 (see Table 4.8). The high share in the level of exports reflects the high

Table 4.7 Trade in the clothing industry (2015–2018)

Description	2015 (Mln/EUR)	2016 (Mln/EUR)	2017 (Mln/EUR)	2018 (Mln/EUR)
Total exports of Albania	1,900	1,902	2,133	2,425
Exports of cloths	363.5	428.5	475	498
Total imports	4,255	4,525	4,892	5,011
Imports of cloths	434.5	504.5	555.6	565.8

Source INSTAT

Table 4.8 Final export destinations of the clothing industry in Albania (2015–2018)

Year	2015			2016			2017			2018		
Description	Total exports (Mln/EUR)	Clothing exports	% of total exports in the clothing industry	Total exports (Mln/EUR)	Clothing exports	% of total exports in the clothing industry	Total exports (Mln/EUR)	Clothing exports	% of total exports in the clothing industry	Total exports (Mln/EUR)	Clothing exports	% of total exports in the clothing industry
Italy (exports of cloths)	966	270	74.28	1,038	312.2	72.86	1,141	342	72.00	1,165	350.3	70.34
Greece (exports of cloths)	74	29.5	8.12	87	36	8.40	91	41	8.63	102	41.2	8.27
Germany (exports of cloths)	59	35	9.63	65	37.4	8.73	85	42.5	8.95	105	45	9.04
Spain (exports of cloths)	98	0.5	0.14	62	1	0.23	117	2	0.42	189	4.8	0.96
France (exports of cloths)	19	8.3	2.28	20	8.5	1.98	23	9.2	1.94	26	9.5	1.91

Source INSTAT

Table 4.9 Production value and value added of surveyed clothing manufacturing enterprises in Albania (2016)

Year	2016				
Description	Production value (Mln/EUR)	Intermediate consumption (Mln/EUR)	Value added in production (Mln/EUR)	Value added/production (%)	Average monthly wage (EUR)
	(1)	(2)	(3) = (1) − (2)	(4) = (3)/(1)	(5)
In the industry	8,500	4,915	3,585	42.18	275
In the manufacturing industry	1,537	1,022	515	33.51	245
In clothing manufacturing	160	72	88	55.00	265

Source Structural Survey of Economic Enterprises (The survey is conducted for an approximate of 80% of the enterprises operating in the year in Albania and for each categorization included in the survey in a given year. Information on them is gathered on varies forms including questionnaires, interviews, information from tax authorities.), INSTAT

number of Italian owned clothing manufacturing enterprises operating in Albania. The second trade partner is Germany with a much lower share at only 9.04% of the exports in the industry in 2018.

In 2016, the share of production in the clothing industry for the enterprises subject to the national survey was 10.6% of the total production[7] generated in the manufacturing industry. This industry in Albania is characterized by a high share of value added in production of 55% (see Table 4.9). Intermediate consumption[8] in the clothing industry is only 51% of the total production and 70% in the manufacturing industry. In 2016, the average monthly wage offered was 265 EUR.

The main cost categories are presented in Table 4.10. Raw materials and personnel costs that together make up more than 85% of the total costs for clothing manufacturers.

In 2016, the clothing industry in Albania is characterized by low level of investments. Investments in the clothing industry are only 5.26% of the total investments undertaken in the same year in the manufacturing industry (see Table 4.11). For the same reporting period, in the clothing industry, the number of enterprises with investments is 12% higher than in the manufacturing industry. Main investments are made in equipment for production and for land where production facilities are located.

---

[7]Production is an activity exercised under the control and responsibility of an institutional unit (enterprise), which combines the sources of labor, capital and raw materials to produce or provide the services.

[8]Intermediate consumption represents value of products or services transformed or totally consumed during the production process. The use of fixed assets is not taken into consideration.

Table 4.10 Production costs of the surveyed clothing manufacturing enterprises in Albania (2016)

Year	2016				
Description	Total costs (Mln/EUR)	Raw materials and consumables (Mln/EUR)	% of total costs	Personnel costs	% of total costs
		Amount (Mln/EUR)		Amount (Mln/EUR)	
In the industry	13,798	5,261	38.13	1,546	11.20
In the manufacturing industry	1,559	1,095	70.21	275	17.65
In clothing manufacturing	151	77	51.02	57	37.96

Source Structural Survey of Economic Enterprises, INSTAT

4.4.3   Integration of the Albanian Clothing Industry into Global Value Chains

The main focus of this section is data analysis in the clothing industry as during the last decade exports in this industry have accounted for one of the largest share of total exports of Albania. The clothing industry is characterized by long production chains including transformation and assembly of intermediate products in more backward linkages in GVC participation [4]. Their results show that export oriented foreign investors are less likely to buy their inputs locally if their country/industry is backwardly involvedin GVCs. As an example, foreign firms in the clothing industry move in locations where it is easier to import and re-export parts and components to third markets. These types of investments are often characterized by minimal levels of linkages with the local economy [26].

4.4.3.1   Data Description

The analysis uses the Inter-Country Input-Output (ICIO) database, World Bank database, Bank of Albania and INSTAT data to construct the time series dataset for the clothing industry in Albania. In addition, this study contributes to the existing literature by using not only an extensive time set of yearly observations from 1990 to 2018, but also by looking at the importance of clothing industry in Albania on the GVC participation.

Our dependent variable is the *gvc participation* in the clothing industry. Borin and Mancini [13] defines *gvc exports* in clothing industry as the value of production crossing more than one border.

$$gvc = gvcb + gvcf$$

Table 4.11 Main investments in the surveyed clothing manufacturing enterprises in Albania (2016)

Year	2016							
Description	% of enterprises with investments among those surveyed	Total investment (Mln/EUR)	Buildings (% of total investment)	Construction and installations (% of total investment)	Machinery and equipment (% of total investment)	Means of transport (% of total investment)	Land (% of total investment)	Other investments (% of total investment)
All market producers	8.60	1,730	42.09	9.10	22.43	4.08	4.06	18.26
Producers in the manufacturing industry	18.10	304	9.89	1.76	30.47	2.55	5.55	49.78
Producers in clothing manufacturing	26.80	16	16.26	4.16	67.68	3.13	3.28	5.44

Source Structural Survey of Economic Enterprises, INSTAT

In the equation, gvcb (global value chains backward) participation in the clothing industry are the aggregate of foreign and domestic value in imported inputs that are re-exported in the clothing industry, while gvcf (global value chains forward) participation in the clothing industry is the value of domestic productions re-exported by bilateral partner countries. Similarly, to the first analysis explained in chapter of this book the determinants for the GVC Participation at the country level, the following indicators are used as independent variables to construct the model at the industry level (Table 4.12).

4.4.3.2 Methodology

The methodology followed in this section is the same as described in Chap. 3, where the determinants of GVC participation at the country level for Albania were analyzed. In this chapter the focus is more at an industry level for Albania, more specifically the clothing industry. Summarized are the steps followed for data set and model analysis are below:

Initially, the descriptive statistics were analyzed for the data set of dependent and independent variables. The descriptive analysis is included in Appendix of this chapter in Table 4.16.

What follows is the construction of the model based on the OLS having the following equation:

$$Y_t = \alpha + \beta X_t + \varepsilon_t$$

- Y_t refers to the dependent variable at time t,
- α is the intercept,
- X_t is the vector of independent variables at time t,
- ε_t is the error term In order to achieve the best fitted model.

In order to obtain the best fitted model, the Dicky-Fuller test is applied as one of the most popular tools to check the stationarity of the variables. Table 4.13 shows the result of Dickey Fuller test for stationarity of selected variables.

In addition, Figs. 4.8 and 4.9 are an example that shows the transformation of variables from non-stationary to stationary (Fig. 4.10). The remaining transformations are included in Appendix of this chapter (see Figs. 4.11, 4.12, 4.13, 4.14, 4.15 and 4.16).

Multicollinearity is another property examined on the time series data sets. It is examined through the correlation matrix between independent variables and a decision is made on the criteria that if correlation is 0.8 or higher in absolute value it is indicative for imperfect multicollinearity. The correlation matrix is presented in Table 4.17 of the Appendix.

The selection of the correct functional model form is based upon examining the scatter plots of the dependent variable and each independent variable, in order to check if there are any non-linear relationships.

Table 4.12 Independent variables, description and the expected impact on the model

Independent variables name	Description	Explanation/source	Expected sign
TextExp_Tot	Exports in the clothing industry to total exports ratio	This variable shows the weight of the clothing industry exports to total exports	+
TextExp_GDP	Export in clothing industry to GDP ratio	Clothing industry exports to GDP ratios are very small. However, it is expected to have a significant effect on the participation of the clothing industry in GVCs	+
FDIsttext_totFDI	FDI stock in the clothing industry to total FDI stock ratio	FDI is the one of the key determinants of GVC participation. However, the variable FDI stock in the clothing industry has a very limited number of observations for Albania	+
lGNI_cap	GNI per capita (USD) is an indicator of the level of development and economic potential	In order to capture the percentage change (elasticity) rather than the unit change, GNI per capita (measured in current USD) is transformed to its logarithmic value. *Source* World Bank	+
unemp	The unemployment rate is an indicator of the level of development and economic potential	It is a lagging indicator, meaning that it generally rises or falls in the wake of changing economic conditions, rather than anticipating them. *Source* World Bank	−

(continued)

Table 4.12 (continued)

Independent variables name	Description	Explanation/source	Expected sign
lNMW	National min wage (USD) is an indicator of the location advantage for attracting foreign direct investments	In order to capture the percentage change (elasticity) rather than the unit change, NMW (measured in current USD) is transformed to its logarithmic value Škabić [71] *Source* World Bank	± (The negative relation is explained by the fact that higher wages lead to lower GVC participation in case of developing countries, while in developed countries that are specialized in high-technology and human capital higher wages will lead to higher GVC participation)
EdExp	Expenditure on education as % of GDP considered as a proxy for the level of absorptive capacity needed to satisfy investors' demand for sophisticated intermediate inputs	Borensztein et al. [12], Amendolagine et al. [4]	+

Table 4.13 The results of the Dickey Fuller test for stationarity in the clothing industry

Variable name	MacKinnon p-value	MacKinnon p-value at first difference
lgvc	0.8414	0.0000***
TextExp_Tot	0.4807	0.0000***
TextExp_GDP	0.2498	0.0000***
FDIsttext_totFDI	0.9988	0.9828
lGNI_cap	0.9038	0.0031***
unemp	0.0023***	
lNMW	0.8237	0.0000***
EdExp	0.005***	

Significant at 1%***, 5%**, 10%*

What follows is checking for the model linearity through the RESET test. It is a general specification test for the linear regression model. More specifically, it tests whether non-linear combinations of the fitted values help explain the dependent variable.

Fig. 4.8 Original time series of log GVC participation in the Albanian clothing industry. *Source* Author's calculation

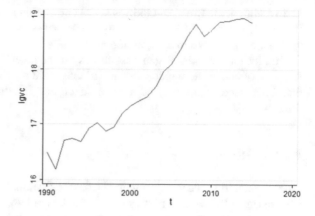

Fig. 4.9 First difference of log GVC participation in the Albanian clothing industry. *Source* Author's calculation

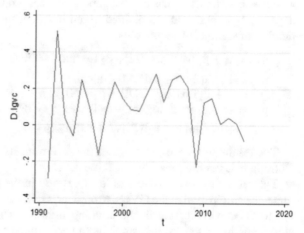

Fig. 4.10 Scatter plot showing relation between the GVC participation of Albanian clothing industry and nominal minimum wage. *Source* Author's calculation

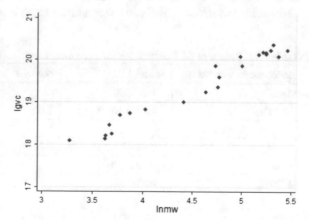

In addition, the Breusch–Godfrey serial correlation LM test is a test for autocorrelation in the errors of the regression model. It makes use of residuals from the model, which are considered in a regression analysis.

Finally, the Breusch–Pagan Test allows for the heteroskedasticity process to be a function of one or more independent variables, and it's usually applied by assuming that heteroskedasticity may be a linear function of all the independent variables in the model.

4.4.3.3   Results

After several trials the results obtained from OLS estimations of the three best models are presented below. The best models are selected based on the highest R-squared, which is a statistical measure that represents the proportion of the variance for a dependent variable that's explained by an independent variable, as well as higher number of significant coefficients.

1. $lgvc = \alpha + \beta_1 TextExp\_Tot + \beta_2 TextExp\_GDP + \beta_3 lGNI\_cap$
$\qquad + \beta_4 unemp + \beta_4 lNMW + \varepsilon$
2. $lgvc = \alpha + \beta_1 TextExp\_GDP + \beta_2 lGNI\_cap + \varepsilon$
3. $lgvc = \alpha + \beta_1 TextExp\_Tot + \beta_2 TextExp\_GDP + \beta_3 lGNI\_cap$
$\qquad + \beta_4 unemp + \beta_5 lNMW + \beta_6 EdExp + \varepsilon$

The results of OLS estimations of the determinants of GVC participation in the clothing industry are presented in Table 4.14.

The results indicate that the ratio of exports in the clothing industry positively impacts participation in the GVCs. Moreover, GNI per capita is positively contributing to GVC participation in the clothing industry. The ratio of FDI stock in the clothing industry to total FDI stock has a very limited number of observations and a descending time trend, which can be seen in the graph in Appendix of this chapter. By

Table 4.14 Results of OLS estimations for three models on the clothing industry in Albania

Independent variable	Model 1	Model 2	Model 3
TextExp_Tot	−1.265309		−3.279512*
TextExp_GDP	4.47e+07***	2.02e+07***	4.06e+07**
FDIsttext_totFDI			
lGNI_cap	0.9486857***	0.6415829***	0.000353**
unemp	−0.0019629		−0.008425
lNMW	0.1589331		0.2163531**
EdExp			−0.176205**
constant	0.0019234	0.0370201	0.1422688
R-squared	0.6619	0.5622	0.7004

Significant at 1%***, 5%**, 10%*

making transformations to make stationary series, many variables are lost. Thus, this variable cannot be used for this model. For future studies more observations for FDI stock and FDI inflow in the clothing industry for Albania are needed. In addition, the minimum wage has a significant impact on participation of Albania in the GVC of the clothing industry, indicating that territories with low minimum wages are attractive for placement of low value added activities withing GVCs. The negative sign of education in the model indicates that the higher investments in education the lower the integration in the clothing GVCs as the country moves into more specialized and highly skilled industries.

4.4.4 Policy on the Clothing Industry in Albania: Façon Package

This section presents the "Façon Package" which constitutes the first set of measures targeting a particular industry introduced by the Government of Albania (GoA) after the establishment of the market economy in the early 1990s.

4.4.4.1 Context of the Façon Package

In 2013, a political rotation occurred in Albania from a right into a left wing government, which changed the system of taxation in the country from a flat system into a progressive one. The new tax system exposes enterprises categorized as "Façon firms" to an increase in the income tax from 10 to 15%. With the aim to ease the tax burden and to maintain the competitiveness of the enterprises in the region, the GoA drafted the "Façon Package" that includes a number of fiscal and economic measures targeting production activity of clothing manufacturing enterprises operating under a Façon regime (inward processing) in Albania.

For the drafting the package Ministry of Economic Development, Trade, and Entrepreneurship established a working group[9] chaired by the Minister was formed and included representatives of the Chamber of Façon of Albania, of clothing and footwear manufacturing enterprises, Chambers of Commerce, a number of line ministries,[10] General Directorate of Customs, General Directorate of Taxation. The working group for this package was set up in the Ministry of Economic Development, Trade and Entrepreneurship with representatives of active Façon enterprises in the country. The working group was continuously supported by a group of academics and experts from the International Center of International Development of Harvard Kennedy School of Government, led by Professor Ricardo Hausmann. The main objective of the working group was to identify the measures to be included

---

[9]The researcher participated in the working group between January–December, 2014.

[10]Ministries involved in the preparation of the Façon Package included: The Ministry of Finance, The Ministry of Social Welfare and Youth, the Ministry of Foreign Affairs.

in the package and derived from an intensive dialogue with Façon manufacturing enterprises.

In its first public declaration, the working group announced the two objectives of the package. The first objective is to increase the value of exports in the clothing and footwear industry by 80%, reaching 900 Mln/EUR and the second objective is to increase every year with 15,000 the number of workers employed in the Façon industry. The Façon Package consists of 32 measures[11] out of which 23 are being implemented. These measures include adoption of 14 legislative acts including 2 laws and 14 governmental decisions.

Financial support to this package is provided through various government funds and donors contributions. According to the Ministry of Economic Development, among the government financed funds are: (i) the Albanian Investment and Development Agency Fund to support the clothing industry with an amount of 360,000 EUR, (ii) the Competitiveness Fund which in 2014 has supported 19 Façon clothing manufacturing firms with a total amount of 200,000 EUR, (iii) the Innovation Fund, which has financed 17 projects in the clothing industry with a total amount of 100,000 EUR, (iii) the Creative Economy Fund has financed 20 projects in the clothing until November 2014 with a total amount of 265,000 EUR or 61% of the annual amount. The objective of the Creative Economy Fund is to finance a fraction of the loans that Façon clothing manufacturing enterprises have acquired in financial intermediaries in order to finance new production lines.

Among the donors, the Italian Cooperation is disbursing of a soft loan with a total value of 40 Mln/EUR for enterprises operating in Albania. By the end of 2015, the amount disbursed to 110 enterprises is 28.5 Mln/EUR. Among the beneficiaries 20 enterprises are clothing manufacturers. The clothing manufacturing enterprises were included as beneficiary enterprises of the Italian Cooperation loan only after the Façon Package became effective. The amounted allocated to clothing manufacturing enterprises by the end of 2015 is approximately 3 Mln/EUR.

4.4.4.2   Main Measures of the Façon Package

The Façon Package was introduced to the public through a wide coverage in the national media. The first measure mostly discussed in the media was on the decrease in the profit tax from 15 to 10% for enterprises operating under the Façon regime so that they can increase the level of retained earnings and make new investments to expand their operational capacity.

The second key measure of the Façon Package regards the VAT exemptions on machinery purchased by enterprises in the Façon industry used in production. Imported machines having an amount of over 500,000 EUR do not pay any VAT. The list of machines subject to VAT exemption increased with implementation of the Façon Package. The enterprises get VAT reimbursement within 30 days of compiling tax declarations. Data from the General Directorate of Taxation shows that between

---

[11]The measures of the Façon Package are included at the end of the thesis.

January 2015 and August 2015 clothing manufacturing enterprises have imported machineries amounting to 2.52 million EUR.

To continue, enterprises subject to Façon Package are subsidized for additional workers. As the third measure, the Ministry of Finance allocated a 1,000,000 EUR reserve fund to be used for Façon enterprises that employed additional workers. For additional workers enterprises would benefit: (i) wages for four months, (ii) one year of social security contributions per worker, and (iii) if the enterprise is a small medium enterprise the government will cover 70% of the training costs.

The forth measure of the Façon Package refers to advantageous loans for Façon enterprises including reduced rates or when the amount of the loan is above 500,000 EUR the collateral required to get the loan is the responsibility of the government. In the Façon Package are also included lease contracts for 1 EUR on existing production facilities. The lease contracts have a duration between 15 and 20 years.

The fifth measure refers to the intention of the Government of Albania to establish the first EPZ in Albania in the Spitalla area,[12] close to the port of Durres that is the largest in Albania. The EPZ has an area of 850 Ha and is expected that one of third of the area will be occupied by clothing manufacturing enterprises producing for international brands that will transfer production from Asian countries. The EPZ will be operational by 2022 and it is expected to create 150,000 new jobs. The Government of Albania has initiated the procedure for the selection of an international developer of the EPZ.

4.4.4.3 Location Advantages/Disadvantages of Albania to Foreign Investors

The package was drafted as Albania provides a number of advantages to attract FDI. This section presents the main advantages and disadvantages of Albania with respect to foreign clothing manufacturing enterprises (see Table 4.15). Depending on which dominate foreign investors decide to locate production in Albania.

Table 4.15 Locational advantages and disadvantages of Albania for foreign investors

Advantages	Disadvantages
Geographical position/proximity to market	Electricity supply problems
Cost of the labor force	Political instability
Government policies to attract FDI	Administrative bureaucracy
Social similarities	Lengthly custom procedures

Source Author's input based on fieldwork

---

[12]The information on Spitalla EPZ was obtained during interviews with government officials in the Ministry of Economy, Trade, Tourism and Entrepreneurship.

4.4.4.4 Location Advantages

The geographical position of Albania is a dominant advantage of the country. Since the early 1990s many Italian, Greek, and German companies have located production in Albania as proximity to the European market ensures a quick response and a faster delivery time of finished goods to end customers [80].

To continue, foreign clothing manufacturing in Albania can benefit from a flexible and young labor force that if properly trained can produce a variety of articles for recognized international brands like Versace, Prada, Diesel, etc. The labor force available in Albania is employed by foreign enterprises at very competitive rates with an average monthly wage of 450 EUR benefiting also from the handy craft skills of the labor force. In addition, foreign investors can outsource at reasonable costs assembly services to [48].

Moreover, the Government of Albania intends to attract more foreign investors in the clothing industry through policies that include a favorable tax treatment for imports on raw materials and machines, reimbursement of the VAT, eliminating transportation fee for Albanian vehicles carrying to Italy final goods in Albania [6] can benefit from improved transport infrastructure on which Government of Albania has heavily invested in recent years. The improved transport infrastructure available in Albania speeds up the delivery of finished articles and eases for the manufacturer compliance with tight delivery schedules [45].

To continue, Albania shares similar social features with many European countries. In particular, Albania shares similar traditions, culture, music, cuisine with Greece and Italy due to immigration and the presence of respective minority groups in the country. For neighboring countries like Italy and Greece the doing business in Albania is facilitated from the language skills locals have in several European languages (English, Italian, Greek, and German) and the presence of foreign investors operating in other industries [47].

4.4.4.5 Location Disadvantages

According to a study of the Ilahi et al. [45] and Kacani [47], foreign clothing manufacturers that decide to locate production in Albania have to cope with several challenges. Production related challenges faced in Albania include frequent interruptions in electricity during production.

Another disadvantage is bureaucracy, which is reflected in slow custom procedures to import raw materials required for manufacturing of goods and export of finished articles. Also, clothing manufacturing enterprises need to cope with bureaucracy in getting permissions required for development of production sites can be slow and arbitrary with requirements fluctuating from one elected government to the other. In addition, foreign clothing manufacturers are exposed to a political instability and economic uncertainty. Political instability is reflected in frequent changes undertaken in governmental reforms like continuous changes in the tax regime that occur when a new government is elected.

Appendix

See Figs. 4.11, 4.12, 4.13, 4.14, 4.15, 4.16 and Tables 4.16, 4.17, 4.18.

Fig. 4.11 Original time
series of log GVC
participation in the Albanian
clothing industry

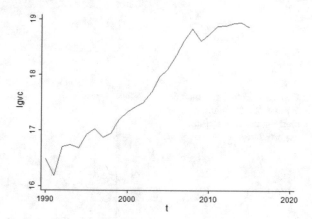

Fig. 4.12 First difference of
log GVC participation in the
Albanian clothing industry

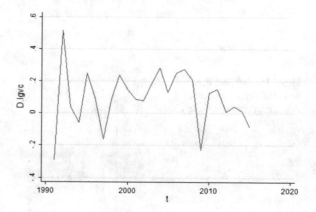

Fig. 4.13 Original time series of exports in the Albanian clothing industry to total exports

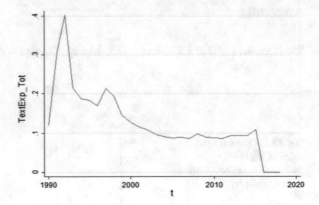

Fig. 4.14 First difference of exports in the Albanian clothing industry to total exports series

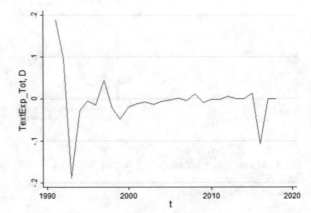

Fig. 4.15 Original time series of exports in the Albanian clothing industry to GDP ratio

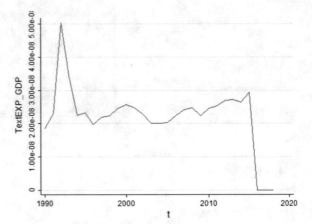

Fig. 4.16 First difference of exports in the Albanian clothing industry to GDP series

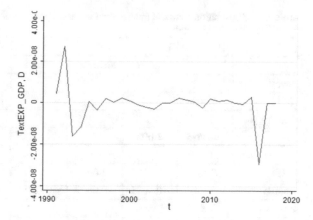

Table 4.16 Descriptive statistics of dependent and independent variables in the Albanian clothing industry

Variable name	Observations	Mean	Standard deviation	Min	Max
lgvc	26	17.74519	0.9229814	16.18933	18.93704
textexp_tot	29	0.1273384	0.0852743	0	0.4040096
textexp_gdp	29	2.23e−08	9.68e−09	0	5.05e−08
fdisttext_~i	7	−0.1132036	3.489875	−7.985357	1.691127
gdpg	29	3.078021	8.188006	−28.00214	13.32233
unemp	29	15.29961	3.437018	9.1	26.5
fdistgdp	27	0.2116256	0.1427837	0.0306666	0.5247372
fdiinfgdp	27	0.0547021	0.0270158	0.0128264	0.0990546
lnmw	24	4.60257	0.7120968	3.273364	5.459586
pt	14	13.02857	4.189259	8.1	20.1
DCredit	25	57.05083	9.047237	38.83821	69.51783
edexp	22	3.489545	0.6208173	2.3	4.5
exp_gdp	29	0.2142772	0.0733633	0.0748482	0.3172197

Table 4.17 Correlation matrix of independent variables possibly contributing to GVC participation in the Albanian clothing industry

	D. lgvc	D. textex-t	D. textex-p	unemp	D. edexp	D. gni_cap	D. lnmw
lgvc							
D1.	1.0000						
textexp_tot							
D1.	−0.4515	1.0000					
textexp_gdp							
D1.	0.1332	0.2255	1.0000				
unemp	−0.0009	−0.4685	−0.0807	1.0000			
edexp							
D1.	−0.2918	0.1585	0.2206	−0.2865	1.0000		
gni_cap							
D1.	0.5943	−0.1424	−0.2358	−0.4206	−0.1119	1.0000	
lnmw							
D1.	0.4275	−0.4993	−0.3785	0.0903	−0.3029	0.3419	1.0000

Table 4.18 Measures of Façon Package in Albania

No.	Business proposals	Government initiatives	Effects
1.	Lowering the price of land and rent of business premises, existing and new investments	1. DCM no. 54 dated 5.02.2014 is adopted 'On defining the criteria, procedure and manner of renting, leasing or other contracts of state property' 1.1 Guidelines 'For determining the criteria, procedure and manner of renting, leasing or other contracts of state property' is adopted	Effect on industry • 1 EUR cost of the lease • About 24 million ALL remain to Façon business every year
2.	Simplification of import-export procedures for secondary compensating products (technological waste)	It adopted Directive no. 1.1, dated 29.01.2014 'On some additions to Instruction no. 01, dated 09.06.2012, On the practical application of the Regime Active processing', article 9, paragraph 2.2.1	Effect on industry • Reduction of administrative costs • Reduction of time for economic operators
3.	Offering custom service at border crossing points even after office hours	Order pursuant to the DCM no. 205, dated 13.04.1999 'On the Customs Code Implementing Provisions', Chapter 4, Article 14 'Working hours of customs offices' has been approved	Effect on industry • 24 h service for declaration and control

4.	The elimination of delays in the reimbursement of VAT	4. Law no. 179/2013 'On tax procedures in the Republic of Albania', amended; has been approved
		4.1 Instruction no. 06 dated 27 02.2014 "On some additions and changes in Instruction no. 17, dated 13.05.2008 'On value added tax' amended", has been approved
		4.2 The DCM no. 202, dated 09.04.2014 'On the determination of terms and conditions to exporters for the purpose of reimbursement of VAT' has been approved
			Effect on industry
			• Refund of VAT will be benefited
			  – Immediately if the taxpayer is a zero risk exporter
			  – Within 30 days if the taxpayer is an exporter
			  – Within 60 days in the case of other taxpayers
			• 900 million leke increase in liquidity for taxpayers

(continued)

Table 4.18 (continued)

No.	Business proposals	Government initiatives	Effects
5.	Clarifying and facilitating business registration procedures (tacit approval)	The efficiency of the implementation of Law no. 9723, dated 05.03.2007 "About the National Center of Registration" will be increased	Effect on industry • Facilitated and improved registration procedures through online application in the portal of Albania • Reduction of recording time from 1 day to 8 h • The platform will be used 20% of businesses within 2014, 40% during 2015, and 50% during 2016 • Attracting online abstracts and papers by 30% of businesses in 2014 and 50% during 2015, 2016
6.	Annual renewal of authorization for fuel consumption	Authorization of VAT on fuel is regulated by Instruction no. 6/3, dated 08.04.2014 "On Amendments to the Instruction no. 17, dated 13.05.2008 'On the value added tax', amended"	Effect on industry • VAT of fuel is immediately credited • Reduction of the time and costs for business
7.	Supply with certificate EUR-1, as a certificate of origin of the Albanian production for all goods exported from Façon entities willing to operate in Albania referring to European standards	Conditions for the granting of preferential origin of goods are determined by the Stabilisation and Association Agreement (SAA)	Effect on industry • Immediate supply with 1 EUR certificate if they meet the defined criteria
8.	Political representation in decision making bodies of Façon industry for the problems of this sector	Associations of Façon industry based on the principle of rotation will be represented in the National Economic Council	Effect on industry • Increase of active dialogue with the government
9.	Strengthening institutional capacities related to formulating support policy for this sector (the creation of an agency, engagement of a contact person, promotion of the image)	Commitment is taken to strengthening the role of AIDA in support of industry. It was agreed the establishment of a counter 'One Stop Shop' dedicated to Façon industry at AIDA	Effect on industry • Financial support for at least six Façon ventures through grant schemes • 10.8 million ALL or 34% of the total grant will go for the Façon industry • Mediation to resolve the problems to be addressed at the counter • Creation of database for this industry

(continued)

Table 4.18 (continued)

No.	Business proposals	Government initiatives	Effects
			• Web page dedicated to the promotion of the sector • Annual conference on Façon from AIDA

10.	Fiscal facilitation and budgetary support from the state for social and health insurance for the new job	Following Amendments were approved
		DCM. 193, dated 02.04.2014 'On Amendments and Additions to the Decision of Council of Ministers no. 47, dated 16.01.2008 of the, "On employment promotion program, through job training", amended'
		DCM. 192, dated 02.04.2014 "On Amendments and Additions to the Decision of the Council of Ministers no. 48 dated 16.01.2008, 'the incentive program for the employment of the unemployed in trouble', amended"
		DCM. 188, dated 02.04.2014 "On Amendments and Additions to Decision of the Council of Ministers no. 199 dated 11.01.2012 'On the extent of funding, criteria and procedures for implementing the program to promote employment of the unemployed workers who enter for the first time at work'"
		CMD Nr. 189, dated 02.04.2014 "On some additions and Amendments to the Decision of the Council of Ministers no. 27, dated 11.01.2012 'On the incentive program for employing women from special groups'"
			Effect on industry
			• 100 million ALL or 3-fold increase compared with 2013 programs to promote employment
			• 1 year social security
			• 4 salaries covered by the state for the categories provided in relevant DCM
			• To promote employment through job training the state finances
			  – 70% of training costs for SMEs
			  – 50% for large enterprises

(continued)

Table 4.18 (continued)

No.	Business proposals	Government initiatives	Effects
11.	Simplification of procedures in the Employment Offices	Ministry of Social Welfare and Youth has simplified procedures in the Employment Offices	Effect on industry • Online application www. puna.gov.al from the subjects • Reduction of the terms of approval of projects from 30 and 20 to 10 days • Performance evaluation based on points for the approval of projects • Fast access to data through the online database for employers and job-seekers
12.	Creating a show room to increase access to international markets. Promoting this sector in domestic and international market	A show room will set be up with donor funding as GIZ (Program for the Development of Enterprises and Employment), Italian Cooperation and USAID, etc.	Effect on industry • Support with technical and logistical assistance for the creation of the show room model
13.	Lack of commercial attachés with the aim of aggressive marketing at the international market	The process to strengthen the role of economic diplomacy has started through establishing the structure of Commercial Attaché, in collaboration with the Ministry of Foreign Affairs	Effect on industry • Promotion of the brand MADE IN ALBANIA • Promoting the image of the country • Increased participation in the wide network of European enterprises Process to strengthen the role of economic diplomacy through establishing the structure of Commercial Attaché, in collaboration with the Ministry of Foreign Affairs
14.	Promoting a pilot industrial zone and promotion of clusters as a new way of attracting foreign investment and increase productivity	Legislation on the function of economic zones and problems for each area declared by the relevant decisions is being revised	Effect on industry • Increase of opportunities for access
15.	Financial support for participation in international fairs and exhibitions	DCM no. 419, dated 05.15.2013, "On the creation of Albanian fund of competitiveness" is being revised	Effect on industry • Increase by 40% of the project value, with a total of 1400.000 ALL

(continued)

Table 4.18 (continued)

No.	Business proposals	Government initiatives	Effects
			• More than 1 application opportunity for this fund • Increasing participation of companies in international fairs and exhibitions
16.	Financial/technical assistance for technology development (information on market, technology and competitiveness of the sector)	It is being negotiated about fundraising • Arrangements for financial programs by donors as Italian Cooperation and KfW (with a value of 39 million EUR)	Effect on industry • Increase of investments for the import of technological lines, through soft loans
17.	Expanding the list of machinery and equipment that are exempt from VAT	In the review process is DCM no. 180, dated 02.13.2013 "On the definition of list of machinery and equipment that are directly related to the investment and exception procedures of the relevant criteria", as amended	Effect on industry • 150 million ALL of business remain exempt from VAT on machinery and equipment
18.	Review the income tax which is considered very high	Income tax return of the Façon industry will be achieved through reallocation in a special fund for employment and training in MMSR	Effect on industry • Employment and training fund is estimated at 5 million dollar value for the first year, which will be gradual • Encouraging employment and workforce training
19.	Eliminating transportation fee for Albanian vehicles towards Italy and Italian plated vehicles that transport on behalf of Façon entities to Albania	It is being consult the signing of an Agreement with the Italian Government for removal of the fixed transport fee	Effect on industry • Reduction of the actual costs € 6.2/t, for 100 km
20.	Increase of efficiency for health care (medical reports for the employed will be once in two years, from currently being once in 6 months)	Regulations are reviewed pursuant to DCM no. 100, dated 03.02.2008 "On determination of hazardous substances"	Effect on industry • Reduce time and costs for industries that are classified as low risk to the health of employees

(continued)

Table 4.18 (continued)

No.	Business proposals	Government initiatives	Effects
21.	Subcontractors will have the same treatment as the main contractor concerning repayment of VAT (VAT discharge to subcontractors)	It is designed the Draft Law "On VAT" which is envisaged to come into force on 1 January 2015	Effect on industry • Reduction of the VAT costs for subcontractors in the amount of about 760 million ALL
22.	Creating special funds for sharing risk and financial resources with donors for the development of exports	Participation in various regional programs funded by the European Commission like COSME, EDIF, etc.	Effect on industry • Stimulation of the Façon industry by providing technical or financial assistance
23.	On line connection of the database of the Tax Directorate with the database of the Social Security Contributions	Project is launched for online connectivity system (single window) of the General Directorate of Taxation with Social Security	Effect on industry • Reduction of time to get from one office to another • Reduction in costs for documentation
24.	Creation of qualified training centers and curricula to serve this industry	In the framework of qualification and vocational training the following have been drafted • ADA project "Dual System Development in Albania" in cooperation with 'Naber Damen Mode' at a cost of 390,000 euros for the establishment of three schools in three cities • Opening a post-secondary course at the Faculty of Mechanical Engineering (textile branch) • Project EU and ILO on the Development of Human Resources (IPA-ILO 2010, HRD)	Effect on industry • Improving the skills of employees • Opening a vocational training course in the field of textile • Certification of knowledge gained outside school (textile industry is chosen as a pilot)
25.	Establishment of vocational centers		
26.	Review of local taxes and fees for the grounds and buildings, where the Façon activity is performed in order to be categorized with the lowest fee or the floor price	In the process of being reviewed is the Law no. 181/2013 'On local tax system', amended	Effect on industry • Establishing good management instruments of this tax, change the model of taxation, from $leke/m^2$ into % of market value

(continued)

Table 4.18 (continued)

No.	Business proposals	Government initiatives	Effects
27.	Forgiveness of interest on arrears for non-payment of social security for the period 1 January 2008–31 March 2014	The Ministry of Finance will draft the law on forgiveness of interest on arrears for social insurance in Façon industry	Effect on industry • Financial impact estimated at about 100 million ALL
28.	Margins should be set for the coefficients applied in the stuff used as raw material, based on the code of the EU customs	Article 140 of the Customs Code specifies the applied coefficient. The new draft Customs Code fully harmonized provisions concerning this regime	Effect on industry • Facilitation of customs procedures
29.	As a reference criterion will be used the amount of ordered final product, as it turns out from customs documentation	Fining business because of the difference in weight is being considered by General Directorate of Customs to address through customs new draft-code	Effect on industry • Facilitation of customs procedures
30.	Silent automatic renewal of authorization for active processing of goods	General Directorate of Customs is working to reflect this issue in the customs new draft-code	Effect on industry • Reduction of financial costs and time for business
31.	Procedures for import/export of goods carried out in selected point by commercial entities	Based on customs legislation, any goods declared at economic customs regimes is excluded from the rule of territorial jurisdiction. The right to establish the mode is enabled at any customs office Customs authority is in a process of preparation for the establishment of legal infrastructure, IT and staff qualification to implement simplified procedures	Effect on industry • Facilitating administrative procedures • Reduce the time for businesses
32.	Authorizations for machines that are temporarily under active processing should not be 1 + 1 year, but be given at customs houses for 5 years	General Directorate of Customs is reviewing procedures to find solutions for 5-year authorization	Effect on industry • Facilitating administrative procedures • Reduce the time for businesses

References

1. Abernathy FH, Volpe A, Weil D (2005) The apparel and textile industries after 2005: prospects and choices. Working paper no. 2005-23. Harvard Center for Textile and Apparel Research, Cambridge, MA
2. Ali M, Habib M (2012) Supply chain management of textile industry: a case study on Bangladesh. Int J Supply Chain Manag 1(2):35–40
3. Altenburg T (2006) Governance patterns in value chains and their development impact. Eur J Dev Res 18(4):498–521
4. Amendolagine V, Presbitero AF, Rabellotti R, Sanfilippo M, Seric A (2017) FDI, global value chains, and local sourcing in developing countries. IMF working paper WP/17/284. Washington, DC
5. Anner M (2019) Squeezing workers' rights in global supply chains: purchasing practices in the Bangladesh garment export sector in comparative perspective. Rev Int Polit Econ. https://doi.org/10.1080/09692290.2019.1625426
6. Atoyan RV, Benedek D, Cabezon E, Cipollone G, Miniane JA, Nguyen N, Petri M, Reinke J, Roaf J (2018) Public infrastructure in the Western Balkans: opportunities and challenges. IMF Publications, Washington, DC
7. Azevedo SG, Prata P, Fazendeiro P (2014) The role of radio frequency identification (RFID) technologies in improving process management and product tracking in the textiles and fashion supply chain. Science Direct 42–69. https://doi.org/10.1533/9780857098115.42
8. Bair J, Gereffi G (2001) Local clusters in global chains: the causes and consequences of export dynamism in Torreon's blue jeans industry. World Dev 29(11):1885–1903
9. Baldwin RE (2011) Trade and industrialization after globalization's 2nd unbundling: how building and joining a supply chain are different and why is matters. National bureau of economic research working paper no. 17716. Cambridge, MA
10. Bamber P, Fernandez-Stark K, Gereffi G, Guinn A (2014) Connecting local producers in developing countries to regional and global value chains: update. OECD trade policy papers, no. 160. OECD Publishing, Paris. https://doi.org/10.1787/5jzb95f1885l-en
11. Belussi F, Sammarra A (2010) Business networks in cluster and industrial districts. The governance of value chain. Routledge, Oxford, United Kingdom
12. Borensztein J, De Gregorio J, Lee W (1998) How does foreign direct investment affect economic growth? J Int Econ 45:115–135
13. Borin A, Mancini M (2019) Measuring what matters in global value chains and value-added trade. Policy research working paper no. WPS 8804, WDR 2020 background paper. World Bank Group, Washington, DC
14. Boström M, Micheletti M (2016) Introducing the sustainability challenge of textiles and clothing. J Consum Policy 39:367–375. https://doi.org/10.1007/s10603-016-9336-6
15. Brewer PC, Speh TW (2000) Using the balanced scorecard to measure supply chain performance. J Bus Logist 21(1):75–91
16. Bruce M, Daly L (2004) Lean or agile? A solution for supply chain management in the textiles and clothing industry. Int J Oper Prod Manag 24(3):151–170
17. Caro F, Martinez-de-Albeniz V (2015) Fast fashion: business model overview and research opportunities. In: Agrawal N, Smith SA (eds) Retail supply chain management: quantitative models and empirical studies, 2nd edn. Springer, New York, pp 237–264
18. Chan FTS (2003) Performance measurement in a supply chain. Int J Adv Manuf Technol 21(1):534–548
19. Choi TM (2016) Information systems for the fashion and apparel industry. Woodhead Publishing Series in Textiles, United Kingdom
20. Christopher M (2000) The agile supply chain: competing in volatile markets. Int J Logist Manag 29(1):37–44
21. Christopher M, Peck H, Towill D (2006) A taxonomy for selecting global supply chain strategies. Int J Logist Manag 17(2):277–287

22. Christopher M, Towill D (2006) Developing market specific supply chain strategies. Int J Logist Manag 13(1):1–14
23. Daspal D (2011) Apparel supply chain and its variants. Mater Manag Rev 7(8):6–9
24. Dicken P (2011) Global shift: mapping the changing contours of the world economy, 6th edn. The Guilford Press, New York
25. Eliiyi DT, Yurtkulu EZ, Sahin DY (2011) Supply chain management in apparel industry: a transshipment problem with time constraints. Tekst Konfeksiyon 2:176–181. Izmir University of Economics, Izmir, Turkey
26. Farole T, Winkler D (2014) Making foreign direct investment work for Sub-Saharan Africa. Local spillovers and competitiveness in global value chains. The World Bank, Washington, DC
27. Fernandes AM, Hillberry RH, Mendoza-Alcántara A (2019) Trade effects of customs reform: evidence from Albania. Paper presented at department of economics spring seminar, University of Nebraska, Lincoln
28. Fernandez-Stark K, Frederick S, Gereffi G (2011) The apparel global value chain, economic upgrading and workforce development. Center on Globalization Governance and Competitiveness, Duke University, Durham, North Carolina, United States
29. Figueiredo PN (2011) The role of dual embeddedness in the innovative performance of MNE subsidiaries: evidence from brazil. J Manage Stud 48(2):417–440. https://doi.org/10.1111/j.1467-6486.2010.00965.x
30. Frederick S, Daly J (2019) Pakistan in the apparel global value chain. Duke Global Value Chains Center, Duke University, Durham, North Carolina, United States
31. Fukunishi T, Goto K, Yamagata T (2013) Aid for trade and value chains in textiles and apparel. OECD/WTO/IDE-JETRO, OECD Publications, Paris, France
32. Gereffi G (1999) International trade and industrial upgrading in the apparel commodity chain. J Int Econ 48:37–70
33. Gereffi G (2011) Global sourcing in US apparel industry. J Text Apparel Technol Manag 2(1):1–5
34. Gereffi G, Fernandez-Stark K, Bamber P, Psilos P, DeStefano J (2011) Meeting upgrading challenge: dynamic workforces for diversified economies. Center on Globalization, Governance and Competitiveness, Duke University, Durham, North Carolina, United States
35. Gereffi G, Memedovic O (2003) The global apparel value chain: what prospects for upgrading by developing countries. Strategic Research and Economics Branch, United Nations Industrial Development Organization, Vienna, Austria
36. Gereffi G, Frederick S (2010) The global apparel value chain, trade and the crisis. Policy research working paper 5281. Development Research Group Trade and Integration Team, World Bank, Washington, DC
37. Gopura S, Payne A, Buys L (2016) Industrial upgrading in the apparel value chain and the role of designer in the transition: comparative analysis of Sri Lanka and Hong Kong. Asia Pac J Multidiscip Res 4(4):103–111
38. Goto K, Endo T (2014) Upgrading, relocating, informalising? Local strategies in the era of globalization: the Thai garment industry. J Contemp Asia 44(1):1–18
39. Goto K, Natsuda K, Thoburn J (2011) Meeting the challenge of China: the Vietnamese garment industry in the post MFA era. Glob Netw 11(3):355–379
40. Grumiller J, Azmeh S, Staritz C, Raza W, Grohs H, Tröster B (2018) Strategies for sustainable upgrading in global value chains: the Tunisian textile and apparel sector. Policy note no. 27/2018. Austrian Foundation for Development Research (ÖFSE), Vienna
41. Humphrey J, Schmitz H (2002) How does insertion in global value chains affect upgrading in industrial clusters. Reg Stud 36(9):1017–1027
42. Hussain D, Figueiredo M, Tereso A, Ferreira F (2011) Textile and clothing supply chain management: use of planning link in the strategic planning process. In: X Congreso Galego de Estatística e Investigación de Operacións, Pontevedra, Spain
43. ILO (2015) Myanmar garment sub-sector value chain analysis international labour organization. ILO Liaison Officer for Myanmar, Geneva

44. ILO (2019) The future of work in textiles, clothing, leather and footwear. Working paper no. 326. Geneva
45. Ilahi N, Khachatryan A, Lindquist W, Nguyen N, Raei F, Rahman J (2019) Lifting growth in the Western Balkans. The role of global value chains and services exports. IMF working paper no. 19/13. Washington, DC
46. Javed A, Atif RM (2019) Global value chain: an analysis of Pakistan's textile sector. SAGE J. https://doi.org/10.1177/0972150918822109
47. Kacani J (2016) Towards knowledge based flexibility for manufacturing enterprises: with a case study. Int J Intell Enterp 4(3):204–226
48. Kacani J, Van Wunnik L (2017) Using upgrading strategy and analytics to provide agility to clothing manufacturing subsidiaries: with a case study. Glob J Flex Syst Manag 18(1):21–31
49. Kaplinsky R, Morris M (2002) A handbook for value chain. International Development Research Centre, Canada
50. Kilduff P, Ting C (2006) Longitudinal pattern of corporate advantage in the textile complex: part 2 sectoral perspectives. J Fash Mark Manag 10(2):150–168
51. Lam JKC, Postle R (2006) Textile and apparel supply chain management in Hong Kong. Int J Cloth Sci Technol 18(4):265–277
52. Li J, Sun W (2009) Study on clothing industry present condition and structure adjustment. Asian Soc Sci 5(10):128–133
53. Lopec-Acevedo G, Robertson R (2012) Sewing success? Employment, wages, and poverty following the end of multi-fiber arrangement, No. 67542, The World Bank, Washington DC
54. Lopez-Acevedo G, Robertson R, Savchenko Y, Kumar A (2016) Stitches to riches? Apparel employment, trade, and economic development in South Asia. The World Bank, Washington, DC
55. Martin M (2013) Creating sustainable apparel value chains: a premier on industry transformation. Impact Econ Prim Ser 2(1):1–41. Geneva, Switzerland
56. Mason-Jones R, Naylor BJ, Towill DR (2000) Engineering the leagile supply chain. Int J Agile Manuf Syst 2(1):54–61
57. Mike M, Barnes J, Kao M (2016) Global value chains, sustainable development, and the apparel industry in Lesotho. International Centre for Trade and Sustainable Development (ICTSD), Geneva
58. Morris M, Staritz C (2017) Industrial upgrading and development in Lesotho's apparel industry: global value chains, foreign direct investment, and market diversification. Oxf Dev Stud 45(3):303–320
59. Morrison A, Pietrobelli C, Rabellotti R (2008) Global value chains and technological capabilities: a framework to study learning and innovation in developing countries. Oxf Dev Stud 36(1):39–58
60. Nordas HK (2004) The global textile and clothing industry post the agreement on textiles and clothing. Discussion paper no. 5. World Trade Organization, Geneva
61. Organization for Economic Cooperation and Development (2008) Defining and strengthening sector specific sources of competitiveness in the Western Balkans, recommendations for a regional investment strategy. Paris, France
62. Organization for Economic and Cooperation Development (2010) South East Europe investment reform index 2010: monitoring policies and institutions for direct investment. Paris, France
63. Perry P, Wood S (2019) Exploring the international fashion supply chain and corporate social responsibility: cost, responsiveness and ethical implications. In: Fernie J, Sparks L (eds) Logistics and retail management, 5th edn. Kogan Page
64. Phu H (2017) Employment and wages rising in Pakistan's garment sector. In: Asia-Pacific garment and footwear sector research note (7):8
65. Ramirez P, Rainbird H (2010) Making the connections: bringing skill formation into global value chain analysis. Work Employ Soc 4:699–710
66. Ray S, Mukherjee P, Mehra M (2016) Upgrading in the Indian garment industry: a study of three clusters. Asian development bank economics working paper series no. 43. https://doi.org/10.2139/ssrn.2941857

67. Rodrik D (2013) The past, present, and future of industrial policy. Global Citizen Foundation, Geneva, Switzerland
68. Romano P, Vinelli A (2004) Quality management in a supply chain perspective. Int J Oper Prod Manag 21(4):446–460
69. Scott AJ (2006) The changing global geography of low technology, labor intensive industry: clothing, footwear, and furniture. World Dev 34(9):1517–1536
70. Scott AJ, Storper M (2007) Regions, globalizations, development. Reg Stud 41(1):579–593
71. Škabić IK (2019) The drivers of global value chain (GVC) participation in EU member states. Econ Res Ekon Istraz 32(1):1204–1218. https://doi.org/10.1080/1331677X.2019.1629978
72. Sturgeon TJ, Memedovic O (2011) Mapping global value chains: intermediate goods, trade, and structural change in the world economy. United Nations Industrial Development Organization, Vienna
73. Taglioni D, Winkler D (2016) Making global value chain work for development. Washington, DC, The World Bank
74. Tanaka K (2018) Assessing the effects of U.S. trade policies on the garment industry in Cambodia. Thail World Econ 36(3):47–65
75. Tewari M (2006) Is price and cost competitiveness enough for apparel firms to gain market share in the world after quotas? A review. Glob Econ J 6(4):1–46
76. Thoburn J (2009) The impact of the world garment recession on the textile and garment industries of Asia. In: Seoul (Korea) workshop, 13–15 Nov. United Nations Industrial Development Organization, Vienna
77. United Nations Industrial Development Organization (2018) Global value chains and industrial development, lessons from China, South-East and South Asia. United Nations Publication, Vienna
78. Van Klaveren M, Tijdens K (2018) Mapping the global garment supply chain. WageIndicator Foundation, 74. https://wageindicator.org/documents/publicationslist/publications-2018/2018
79. World Bank (2009) Building competitiveness in albania, europe and central asia region. No. 47866-Al, Washington DC
80. World Bank (2019) Rising uncertainties. Western Balkans regular economic report no. 16. Washington, DC
81. Yaozhong L (2016) Promoting the participation of SMEs in textile and apparel global value chains in China. Asia-Pacific Economic Cooperation (APEC)

Chapter 5
A Data Centric Approach on Case Study Methodology in the Clothing Manufacturing Industry

Abstract The case study methodology introduced in this chapter supplements the data centric approach on global value chains. The objective of this chapter is to introduce the characteristics of case study methodology, to consider its advantages and limitations, to mention applications of case study methodology in the literature, and to present the steps followed in analyzing the case studies. The case study analysis provides additional insights on the degree of integration of clothing manufacturing subsidiaries into global value chains.

Keywords Data centric research framework · Case study methodology · Internal validity · External validity

5.1 Introduction

Case study research is one of the methods used by social science researchers [22, 42]. Initially, case studies were developed from systematic fieldwork and investigations of contemporary phenomena through observations of the researcher and key information obtained from people involved and affected by the phenomena of interest [36, 44]. The approach of case study methodology combining fieldwork and numerical information in the areas of interest was promoted by researchers of the University of Chicago that practised field study work on the contemporary society in the surroundings of universities [25, 26]. In case study methodology the researcher gets to know his area of interest by understanding relevant processes, operations, and the events for which information was obtained from a wide range of sources including direct observation and face to face interviews.

The literature has identified the following main characteristics of case study methodology [9, 15, 42]:

- **It is based on an evolving design**.
  The design of case study research cannot be fully specified in advance of fieldwork. The design develops and evolves based on the analysis generated from the information and data obtained during fieldwork.

- **It is descriptive**.
  A key part of case study research is describing and understanding the dynamics of the phenomenon. Description includes a detailed account of the context, the activities, the participants, and the processes of the phenomena under investigation.
- **It is mainly concerned with understanding the dynamics**.
  Case study is primarily concerned with understanding and describing dynamic and complex processes characterizing the area of interest.
- **It involves fieldwork**.
  Fieldwork implies that the researcher has direct and personal contact with people involved in a phenomenon in the natural setting of the phenomenon.
- **The researcher is the primary instrument for data collection and analysis**.
  The researcher collects the data through questionnaires, surveys, interviews, direct observation and other data collection instruments.
- **It has an inductive approach**.
  Qualitative research is based on inductive logic. As such, it aims to draw conclusions through a bottom-up analysis of specific observations units representing the phenomena under investigation. On the other hand, deductive logic is a top-down going from the general to specific observation to draw specific conclusions [35, 41].

In the literature, there are various approaches on case study methodology depending on the scientific discipline. However, in this chapter a more general version of case study is considered. For Yin [47], a case study is an empirical examination that explores a contemporary phenomenon within its real life context. He states that the distinctive need for case study research comes from the desire to comprehend complex social phenomena. For Tsang [43] and Stake [39], a case study is useful when the opportunity to learn is of primary importance. They regard case study as an intensive study of a single unit for the purpose of understanding a larger group of similar units. A unit refers to a bounded phenomenon e.g., a nation-state, revolution, political party, election, or person-observed at a single point in time or over some delimited period of time. To summarize, both Tsang [43] and Stake [39], consider case study methodology as a mode of inquiry for an in depth examination of an area of interest and phenomena that are part of it.

5.2   Reasons for Case Study Methodology

Ridder [37] and Yin [47] advocate that case study methodology is appropriate when the researcher examines a phenomenon that occurs in present time. This is a peculiarity of case study methodology as data collection methods like interviews and direct observation are inappropriate to study events occurred in the past. They continue by arguing that case study methodology facilitates exploration of a phenomenon by using a variety of data sources. This ensures that the issue is not explored through one dimension, but rather through multiple dimensions that allow the researcher to obtain different perspectives of the phenomenon that is examined. According to Bian et al. [3] and Yin [47], a case study methodology should be considered when: (i) the focus of the study is to answer "how" and "why" questions and (ii) the behavior of those involved in the study cannot be altered.

In addition, Osano and Koine [31] and Flyvbjerg [13] suggest that case study has a level of flexibility (no limitations in the variables included, number of people that can be interviewed, questions to be asked, etc.) which is not offered by other research methods. It also involves careful and in-depth consideration of the case, historical background, physical setting, and other institutional and political contextual factors. Creswell [8] argues that a good case study research relies on a well-constructed validity. Internal validity refers to establishing casual relationships on a given set of conditions. A researcher investigator has to infer that a particular event resulted from some earlier occurrence. He suggests that in case study research casual relationships are well established due to the researcher's direct exposure to the events of interest. The main advantages of case study methodology are (i) to see is to understand better, (ii) to take into account many variables, and (iii) to get feedback during investigation resulting from flexibility in case study [22].

5.2.1 Direct Observation

In a case study, the researcher engages in direct observation. Helper [18, p. 228] states, "you can observe a lot just by watching". Direct observation helps the researcher to take a variety of roles while participating in the phenomena under investigation. This helps the researcher to perceive the reality from the viewpoint of someone inside rather than external to the phenomena. If the researcher is offered the opportunity, he can participate in meetings and discussions with various stakeholders. The reliability of observational evidence increases with multiple observations of phenomena. That is why the researcher observes the phenomena several times. In the same line, Miles et al. [27] and Helper [18] argue that in case study active observation requires a lot of concentration and a high level of objectivity for the researcher to maintain the sequence of events, to obtain the complete picture, and to prevent him from drawing conclusions based on its own beliefs. Direct observation, maximizes the ability of the researcher to understand the perceptions of the subjects involved, to understand the phenomenon in its natural environment, and to build knowledge from his own observations and from the subjects involved. In addition, there are various forms of participant observation [5]. The researcher may be a full participant in the events being examined, purely an observer, or being somewhere in between. The participant understands the events as an insider can bring a particular kind of interpretation to the data obtained from documentary evidence [8].

5.2.2 Consideration of Many Variables

Data collection in empirical and statistical research methods is based primarily on numerical observations obtained from institutions in charge for publishing time series and cross section data. Differently, in a case study methodology the researcher can collect data and information from a variety of sources including documents, archival records, and interviews [9, 20]. Various sources of data and information help the researcher to consider alternative hypothesis and outcomes not identified before the data collection process. Each data source has particular advantages and disadvantages, and by using a combination of sources and techniques, scarcities of one source are supplemented by the advantages of another source. Moreover, Miles [26] argues that in case study the researcher has the opportunity to correctly establish casual relationships among the variables under investigation (internal validity). This methodology helps the researcher to concentrate on the most important issues of the research and to propose appropriate policy recommendations.

To continue, in a case study research consideration of many variables results also from combining qualitative with quantitative evidence that benefits the researcher [40, 45]. Quantitative evidence obtained through various sources can indicate relationships among the variables of interest that have not been previously identified by the researcher [22, 25]. Quantitative evidence in case study keeps the researcher from being carried away by impressions and casual relationships obtained from analyzing qualitative data [40, 43]. Qualitative data is useful to the researcher for understanding the rationale or the theory underlying the relationships revealed from a quantitative analysis or to suggest new theoretical implications that can be strengthened by quantitative support [11, 37].

5.2.3 Flexibility and Feedback

Flexibility in case study methodology, offers the researcher the opportunity to discover facts on its own, to confirm or deny them during direct observation and interviews, and to introduce new propositions that the researcher did not think about. A good way to obtain feedback in a case study research is to conduct face-to-face interviews. Interviews are an essential source of information as a number of prepared and on the spot questions can be addressed. A researcher can conduct two types of interviews [3, 47]. The first one is an open-ended interview. It is very casual and does not comply with a fixed set of questions. In an open-ended interview, a researcher can combine specific and general questions to ensure that important issues are discussed or to allow space for the respondent to suggest topics. During an open-ended interview, the respondents can provide new insights on the matter or recommend new sources of evidence. A drawback of an open-ended interview is that it can be time consuming and may not generate the type of data that is needed for case study analysis. The second type is a focused interview that is designed for a short period and is likely to follow a set of prepared questions. For effective interviews, good preparation prior to the interview will most likely make it more interesting to the respondent and is perhaps the key to productive interviews [30, 41]. In an interview, it is often useful to push the respondent to justify a particular interpretation with further arguments and details. This helps to ensure the accuracy of the information acquired. In addition, active and serious participation in an interview means to build problems for respondents in the interview so that people interviewed can see themselves as problem solvers. This is a trick to get a more active participation from people interviewed [27, 38].

5.2.4   Limitations of Case Study Methodology

According to Bian et al. [3] case study methodology is mainly questioned on generalization for a broader population (external validity). Another threat to case study methodology is the influence of researcher's previous experience on the area of interest. The researcher needs to start a case study without any personal bias in finding out on the phenomenon under investigation. Pre-existing assumptions and experience can impede the researcher from observing important and hidden details in the phenomenon. Such biases on the researcher damage the credibility of the data collection process and the overall trustworthiness of the case study research. Creswell [8] together with Guba and Lincoln [17], suggest that the researcher should strive for a stance of unbias neutrality that can lead to data that is reliable, factual, and credible. In a case study methodology, the researcher will be affected by what is being studied. The repetitive nature of data collection, analysis, and synthesis implies that the researcher is continually learning and understanding more about the phenomenon. Consequently, as the primary instrument for data collection and analysis, the researcher needs to change over the course of the research to reflect the on-going findings and not to stick to previous experience or assumptions [14, 41].

5.3   Steps in Case Study Methodology

In a case study methodology, the sample can be categorized as (i) purposeful, (ii) criterion based, and (iii) theory based sampling. Purposeful sampling aims at selecting cases rich in information from which the researcher can learn a great deal on the issues and events of key importance to the research objective [20, 33]. The logic behind the criterion sampling is to select cases that meet a predetermined set of criteria to help the researcher in answering his questions. Finally, theory based cases assure the researcher to base its conclusions on relevant theory. After categorizing the sample, the steps followed in case study research, according to Tsang [43] and Baxter and Jack [1] (see Fig. 5.1) are:

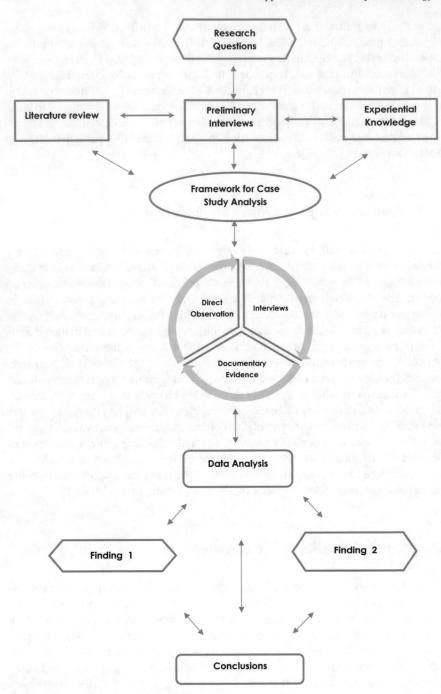

Fig. 5.1 Steps in case study methodology in clothing manufacturing subsidiaries. *Source* Author's drawing based on Tsang [43], Baxter and Jack [1]

Preliminary research

The main activities performed by the researcher consists in: reviewing the literature, conducting initial interviews with experts in the field, and identifying the professional experience relevant to research objectives. Preliminary research not only widens the knowledge of the researcher but also brings to his attention contrasting views.

The framework

Based on the information obtained in the first step, the researcher prepares the framework in which he shapes the phenomenon of interest together with corresponding research questions in line with the overall objective of the research.

Fieldwork

After finalizing the first two steps of case study methodology, the researcher starts fieldwork during which information and data is collected by using multiple techniques including direct observations, interviews, and documentary evidence.

Data analysis and conclusions

After finalizing the fieldwork, the researcher systematically reviews the information and the data collected by using various data analyzing techniques. From the analysis, the researcher confirms or rejects previously identified research questions during the preparation of the framework. As a final step, the researcher draws overall conclusions on the area of interest and prepares policy recommendations.

5.3.1 Ensuring Quality in Case Study Research

A central concern in case study methodology is to integrate different techniques to assure the quality and to generate trust in the findings of the research. Quality in case study research has three components that include trustworthiness, validity, and reliability. Referring to Denzin and Lincoln [9], as a foundation to obtain credibility researchers need to: (i) clearly define research objectives, (ii) carefully formulate purposeful sampling strategies, (iii) systematically collect and analyze data and information, (iv) correctly analyze the data and the information obtained during fieldwork.

Miles et al. [27], Yin [47] and Golafshani [16] argue that validity and reliability are key determinants in a case study of good quality (see Fig. 5.2). A case study research is reliable if it has internal and external validity.

As previously mentioned, validity refers is taking appropriate steps in order to study the phenomenon of interest, for Creswell [8]. Validity is of two kinds internal and external:

Fig. 5.2 Validity and
reliability in case study
methodology. *Source* Author
representation based on
Miles et al. [27] and Yin [47]

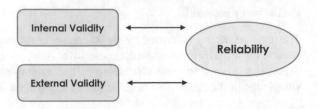

Internal validity

It refers to establishing causal relationships among the variables taken into account
in a case study research. By relating the variables with each other, the researcher
makes interference on the phenomena under consideration.

External validity

It refers to generalizing the findings of a case study to greater groups than those
analyzed in a case study. Generalization is not automatic, as it should be based on
theory. With reference to generalizing from case studies Killam [22], Yin [47] and
Guba and Lincoln [17] argue that theory in case studies is not only of immense
aid in defying appropriate research design and data collection but also becomes
the main instrument for generalizing and ensuring external validity in case study
methodology. Theory is key. It is the only way to begin to generalize from case
studies. Even though generalizing is not easy, selection and repetition of multiple
case studies help to generalize. The use of multiple sources of evidence help the
researcher to strengthen its validity. In addition, reliability reinforces the case study
validity. The ultimate goal of reliability is to minimize errors and biases in a study
and to strengthen its external validity [5].

In addition, Riddler [37] and Yin [47] suggest that placing boundaries on what
elements of the case study are considered for analysis ensures that it will remain
reasonable in scope. The establishment of boundaries in case study research is similar
to the development of criteria for sample selection. The difference between the two
methods is that by placing boundaries in a case study the researcher indicates the
breadth and the depth of the study and not simply the sample of the research. For
Creswell [8], in a case study methodology boundaries are determined by: (i) time
and place, (ii) time and activity, and (iii) definition and context of the case [7].

Reliability

It aims to minimize the errors and biases in a case study by keeping track of all
persons and data collected as if external reviewers will perform an "audit".

According to Miles [26] and Golafshani [16], trustworthiness in case study
research depends on several criteria that are mentioned in Table 5.1.

Table 5.1 Criteria for a case study with a good degree of quality

No.	Criteria for trustworthiness	Usefulness
1.	Credibility	To assure value in the findings of a case study methodology by documenting all steps followed during the research
2.	Transferability	To assure that steps and techniques followed in data collection and interpretation can be applied by other or similar research
3.	Dependability	To assure proper documentation of sources of information when required for external inspection
4.	Confirmability	To assure that the researcher is neutral so that the findings derive only from the information obtained by using multiple sources of data and do not include any bias and interests of the researcher
5.	Prolonged engagement	To assure that the researcher was involved long enough in order to enable the researcher to understand better the characteristics and peculiarities of the phenomena under consideration
6.	Inclusion of primary source material	To assure that information relevant to the research objectives is included in analysis of case studies so that the researcher does not drift away from research objectives
7.	Purposeful sampling	To assure that sufficient data from a sample of units under investigation that meet the overall purpose of the research so that the researcher can draw unbiased conclusions

Source Author representation based on Mills [26] and Golafshani [16]

5.3.2 Case Study Methodology in the Literature on Inward FDI and the Clothing Industry

Case study research has and is widely accepted as an effective methodology to investigate on the impact of FDI in host territories. Blomstrom and Kokko [4] argue that initially FDI impact was examined in 1960s using case studies, while the theoretical arguments appeared only in late 1970s. For example, Johanson and Wiedersheim [19] analyze the effects of embeddedness of four Swedish companies (Sandvik, Atlas Copco, Facit and Volvo). Behrman and Wallender [2], examine the operations of General Motors, ITT, and Pfizer in several host countries focusing mainly on backward linkages. Evan [12] studied the impact of FDI in the Brazilian textile industry while Langdon [23] investigated foreign subsidiaries in the Kenyan soap industry. In addition, Munday et al. [28] use case study methodology to examine the impact of Japanese manufacturing companies in Wales. Phelps et al. [34], rely on a case study methodology to evaluate the impact of LG green field investment in South Wales. Additional studies are those of Larrain et al. [24], who investigate the impact of Intel (manufacture of microprocessors) in Costa Rica. To continue, Nayak [29] looks into Suzuki Motor Corporation as a model of inward FDI in an emerging economy like India. Moreover, Paprzycki [32] uses case study methodology to study the impact of FDI in Japan. He selects subsidiaries from the automobile, finance, and health care industries. Dikkaya and Keles [10] choose two Turkish subsidiaries and a Japanese subsidiary to evaluate the impact of FDI in Kyrgystan. Callychurn et al. [6], highlight the mechanism used by a successful Mauritian clothing manufacturing company to sustain its success despite operating in a difficult and competitive business environment while Wen et al. [46], study the new product development adopted as a competitiveness factor in a Taiwanese textile and clothing manufacturing company. Kacani [21] studies clothing manufacturing subsidiaries in Albania and investigates on their upgrading in the host territory.

5.4   Case Study Methodology on FDI in the Clothing Industry in Albania

This section presents the case study methodology implemented with regard to FDI in the clothing industry in Albania. It starts with the main reasons of why this methodology is applied and it continues with the main steps followed to analyze four clothing manufacturing subsidiaries operating in Albania.

5.4.1   Reasons for Case Study Methodology in Albania

Yin [47], argues that case study research permits the researcher to explore areas with little pre-existing numerical data. Research on FDI in clothing manufacturing industry in Albania, is a recent phenomena that began in the early 1990s and intensified only after 2000s with the establishment of the market economy and recovery of the country from the civil war in 1997. Currently, the data available on FDI in Albania and more specifically those for the clothing industry are very limited and comes mostly from the information provided by the Bank of Albania. Since the establishment of the market economy, only two reports on FDI in Albania are published. United Nations Development Programme published together with the Government of Albania the first report on FDI in Albania only in 2010 and the second report in 2011. Currently, little available numerical data (time series, cross section, etc.) that consists in a sample of less than ten observations on the overall stock of FDI in Albania was firstly estimated only in 2007, leaving little room to use empirical methods like regression analysis or forecasting. Another reason why case study is the primary methodology of this research is that the literature consists mostly of various donor's reports that introduce descriptive analysis of FDI trends in the country. These reports provide policy recommendations on foreign investment that cover the overall business environment in Albania including a broad range of industries, without focusing on a particular sector and without examining the numerous variables required for a thorough investigation of a particular industry. Contrary to the existing literature, the case study methodology in this research aims to yield policy recommendations on foreign enterprises in the clothing industry in Albania based on a thorough investigation of the variables included in the proposed framework.

To continue, in the proposed framework the impact of manufacturing FDI depends on many factors. It is very difficult to include all explanatory variables of the framework into a statistical/econometric model. In such models, only a limited number of variables part of the framework can be tested empirically [4]. Because of the limited number of variables that can be included in empirical models and of the vast data requirements, findings derived from statistical/econometric models may have limited explanatory power and may capture only a fraction of the overall FDI impact. However, in a case study research the researcher can consider many more variables of interest than the available data [20, 47]. Moreover, a case study methodology offers the opportunity to observe how enterprises operate on a daily basis. The literature does not account for the procedures and steps followed by subsidiaries in production processes and especially on how technology is transferred, how raw materials and intermediary inputs are obtained from local suppliers, how training programs have proved beneficial for employees, and how do subsidiaries upgrade [46].

The four case studies can be categorized as descriptive and representative. They are descriptive as their activity since the beginnings of operating in Albania up to 2015 is described. The cases studies in this research are also representative as the objective of a representative case study is to examine the circumstances and conditions of an "ordinary unit of analysis" so that it can help in generalizations. Investigating on qualitative effects of foreign clothing enterprises in a host territory is considered as a common unit of analysis in that particular industry.

5.4.2   Phases Followed in Analyzing Case Studies from the Clothing Industry in Albania

In order to analyze the case studies of this research that consist in four clothing manufacturing subsidiaries five steps were undertaken. The steps are summarized in Table 5.2.

Phase 1—Preliminary research

Preliminary research focused on a comprehensive literature review: (i) on the effects of FDI in host territories, (ii) on the characteristics and the main players in the clothing industry, and (iii) on FDI investments and the clothing industry in Albania. What followed the literature review was the identification of institutions and organizations

Table 5.2 Sources of information in case study research in the Albanian clothing industry

No.	Category	Type of data	Purpose
1.	Material and documentary evidence	Official records on the establishment of the subsidiary and subcontractors Historical data on the operating activity of the subsidiary, subcontractors, owners, suppliers Financial records on the activity of the subsidiary (confidential information) Administrative procedures on the operations of the subsidiary	Useful in grasping the activities, of owners subcontractors, workers, and suppliers of each subsidiary and to obtain necessary information on the performance from the beginning of the operational activity in Albania
2.	Interviews	Textual data from conversations held with the interviewers Historical information obtained during the conversation with representatives of institutions and organizations involved Information on the establishment of additional production plans in the subsidiaries Information on the government policies to attract FDI in general and in the clothing industry in particular Information on the difficulties encountered by clothing manufacturers in Albania Information on the trends of the clothing industry in Albania	Useful information for analyzing the four case studies according to the proposed framework The information obtained is necessary to draw overall conclusions in accordance with research objectives
3.	Participant observation	Textual data in the form of field notes Descriptions of the events, activities and processes experienced by the participant	Useful information in collecting descriptive details about the cases and helpful to interpret and understand the data collected during interviews and those obtained from documentary evidence
4.	Statistical information	Numerical data mostly macro-economic in nature (trade, foreign investment, employment, number of enterprises, GDP etc.), clothing industry and FDI in Albania	Useful information in understanding and analyzing the economy, clothing industry, and foreign investment in Albania. Useful for establishing links between the four case studies and overall business environment

Source Author's design based on fieldwork

to consult and the preparation of the list with representatives to contact. In addition, the main sources of numerical data were identified. Main sources of data include the National Statistical Institute (INSTAT), the Bank of Albania, the Ministry of Finance, the World Bank, and the Chamber of Façon of Albania. By looking into the type and the categorization of the accessible data, it was possible to pinpoint what important numerical information on the clothing industry in Albania was available and not available after the fall of communist regime in the early 1990s. Finally, in this phase a draft questionnaire for the foreign clothing manufacturing enterprises was prepared and was consulted with the representatives of the Chamber of Façon in Albania and the Textile Department of the Polytechnic University of Tirana.

Phase 2—Preparation of the framework

The second phase was dedicated to the preparation of the framework used to analyze the clothing manufacturing enterprises selected for this research. The framework is based on two main pillars that are the knowledge transferred in the host territory and evolution in the quality of the subsidiary. In addition, for each pillar respective sub-areas were identified. After identification of the pillars with the corresponding sub-areas, the graphical representation of the framework was prepared. The framework was discussed and finalized after consultations with representatives of the Chamber of Façon in Albania and the Textile Department in the Polytechnic University of Tirana. Upon finalization of the framework, the researcher set out the criteria for selecting foreign clothing manufacturing enterprises operating in Albania. Finally, during this phase, the questionnaire on the enterprises was finalized and a tentative schedule of meetings with stakeholders was prepared.

Phase 3—Fieldwork

The third phase of the research was dedicated to fieldwork a key prerequisite to analyze the foreign clothing manufacturing enterprises in Albania and to interview institutions and organizations identified during phase the first and the second phase. The criteria used to select the four cases studies in this research were:

- subsidiaries that started the operational activity in the early 1990s, ensuring this way that the enterprises have been manufacturing in Albania for more than 15 years;
- subsidiaries that are geographically spread in the main regions of Albania where the clothing industry is present;
- subsidiaries representing the nationality of the main export destinations of finished articles in the clothing industry;
- subsidiaries that would cover all the variety of goods manufactured in Albania (underwear, sport outfits, traditional dresses, dresses, trousers, swimwear, etc.).

The fieldwork can be characterized both as intensive and extensive. Numerous efforts spent on identifying and contacting relevant individuals in order to obtain accurate information in accordance with research objectives set out in the first chapter of the thesis. Institutions and organizations in this research fall into three main categories as presented in Fig. 5.3. The first category refers to institutions contacted

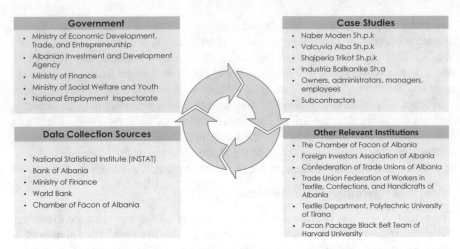

Government

- Ministry of Economic Development, Trade, and Entrepreneurship
- Albanian Investment and Development Agency
- Ministry of Finance
- Ministry of Social Welfare and Youth
- National Employment Inspectorate

Case Studies

- Naber Moden Sh.p.k
- Valcuvia Alba Sh.p.k
- Shqiperia Trikot Sh.p.k
- Industria Ballkanike Sh.a
- Owners, administrators, managers, employees
- Subcontractors

Data Collection Sources

- National Statistical Institute (INSTAT)
- Bank of Albania
- Ministry of Finance
- World Bank
- Chamber of Facon of Albania

Other Relevant Institutions

- The Chamber of Facon of Albania
- Foreign Investors Association of Albania
- Confederation of Trade Unions of Albania
- Trade Union Federation of Workers in Textile, Confections, and Handicrafts of Albania
- Textile Department, Polytechnic University of Tirana
- Facon Package Black Belt Team of Harvard University

Fig. 5.3 Relevant institutions and organization contacted during fieldwork in Albania. *Source* Author's drawing based on fieldwork

to obtain numerical (quantitative information). From these institutions numerical information on the regional economy, FDI in Albania, the clothing industry, and the doing business environment in the countries located in WB countries was obtained. The second category refers to people interviewed who were helpful in providing information on the clothing industry, people from the government, and those in institutions that are concerned with attracting FDI. The third category refers to the group of people that are relevant in the operational activity of clothing manufacturing subsidiaries. From these categories, it was possible to obtain information on different areas covered in this research. In order to receive the needed information frequent contacts were kept with all institutions and organizations.

During this phase, a special initiative was undertaken with the Chamber of Façon in Albania, which aimed at identifying the total number of clothing manufacturers in Albania. This information was not available in any Albanian institutions, that is in charge to publish official statistics. The Chamber of Façon in Albania started to identify all clothing manufactures (foreign and domestic) spread throughout Albania. To continue, interviews with additional relevant organizations were conducted on an ongoing basis. Such organizations include the Foreign Investors Association of Albania, Confederation of Trade Unions, Polytechnic University, donors that have provided funds for strengthening the clothing industry in Albania, and representatives of Harvard University, Center for International Development, involved in the preparation of the Façon Package in Albania.

Questions addressed were according to a predetermined questionnaire and supplemented with on the spot questions depending on the information that was generated. On-site visits were useful to the researcher to observe operations in all departments and units of the subsidiaries and to understand better the organization and the level of dependency from the head office. The researcher interviewed different levels of employees ranging from administrators, to middle management, and assembly work-ers. The researcher also talked with representatives of all local subcontractors work-ing with the clothing manufacturing enterprises under consideration while on-site visits were realized only in three subcontractors.

Frequent contacts with relevant institutions and organizations ensured a cor-rect identification of various categories of sources of information from which the researcher was able to generate different types of data. Table 5.3 mentions the main sources of information, the data, and the purpose of the information for each category.

Phase 4—Analysis

After collecting the necessary information and finalizing the fieldwork, the analysis on the four case studies was prepared. The analysis was developed based on the comparison among them according to the two main areas of the framework. The comparison served to identify the similarities and the differences between the four cases. In the analysis, an ongoing reference is made to rival and supportive supposi-tions on the qualitative effects of FDI in the host territory. In addition, in this phase the researcher identified limitations encountered in the activity of foreign clothing manufacturing enterprises in the local economy. Finally, the analysis on the four case studies was discussed with representatives of the Chamber of Façon of Albania, academics of the Polytechnic University of Tirana, the University of Tirana.

Phase 5—Findings, conclusions and policy recommendations

The last stage of the research was to prepare the findings on the four case studies based on the framework under which the analysis was performed. Identification of findings, served to draw general conclusions on the qualitative effects of foreign clothing manufacturing enterprises in Albania. Based on the findings and conclusions the researcher identified possible policies to propose remedies on the gaps identified during the analysis, on upgrading the clothing industry in Albania, on possible policy recommendations on making more competitive the clothing industry in Albania, and on areas of further research.

5.4.3   Establishing Reliability in Case Study Methodology

The researcher tried to establish trustworthiness in case study methodology by fol-lowing the criteria previously introduced for case study research and more specifically as presented in Table 5.4.

Internal validity

It occurs by establishing relationships between numerous variables identified and obtained for each case study. In order to compare the four case studies the same variables were considered for each case so that similarities and differences could be

Table 5.3 Main phases in case study methodology in analyzing the four clothing manufacturing subsidiaries in Albania

	Phase 1	Phase 2	Phase 3	Phase 4	Phase 5
	Preliminary research	Preparation of the framework	Fieldwork	Analysis	Conclusions and policy recommendations
Goal	Preliminary research to obtain knowledge on the research area in order to identify the steps to follow in accordance with the overall objectives of the book	Preparation of the framework to identify the main areas in order to analyze the four clothing manufacturing subsidiaries	Fieldwork to obtain the necessary data and information from a variety of sources required to analyze the four clothing manufacturing subsidiaries	Analysis in order to find the similarities and differences between the four cases. The analysis serves also as the basis for drawing conclusions on the research	Drawing of conclusions to respond to research objectives and to identify areas of further research
Activities	Extensive review of the literature on the effects of FDI in host territories Review of the literature on the characteristics of the clothing industry	Classification of qualitative effects to be included in the framework Identification of two pillars of the framework consisting in:	Interviews with government officials (deputy minister, secretary general, directors, head of sectors, and experts) involved in the clothing industry in Albania Interviews with local government authorities in the four regions where subsidiaries are located	Comparison between the four subsidiaries according to the framework prepared in phase 2 Identification of similarities and differences between the four cases	Drawing of conclusions for each pillar of the framework Drawing of general conclusions on qualitative effects of foreign investment in the clothing industry in Albania

(continued)

Table 5.3 (continued)

Phase 1	Phase 2	Phase 3	Phase 4	Phase 5
Preliminary research	Preparation of the framework	Fieldwork	Analysis	Conclusions and policy recommendations
Review of the literature on foreign investment and the clothing industry in Albania Identification of the type of data from sources like: (i) INSTAT; (ii) Bank of Albania; (iii) Ministry of Finance; (iv) additional sources Identification of institutions and organizations relevant to the research area Identification of persons to contact on behalf of the stakeholders involved in the clothing industry in Albania,	(i) knowledge transferred in the host territory; (ii) evolution in the quality of the subsidiary; Identification of subareas to be included in each pillar of the framework. Preparation of the graphical representation of the framework Discussion and feedback on the framework with Chamber of Façon in Albania and other relevant stakeholders, determination of the criteria used to select the foreign clothing enterprises in Albania	Interviews within the four subsidiaries (clothing manufacturers) under investigation. The persons interviewed in the subsidiaries include: (i) administration/high management; (ii) middle management; (iii) workers offering mostly assembly services Interviews and on-site visits to subcontractors of subsidiaries	On-going reference to rival and supportive propositions identified from literature review during phase 1 and phase 2 of the research Continuous reference to numerical data collected on subsidiaries and to those related to the local economy, foreign investment, and clothing industry in Albania Identification of limitations in the qualitative effects of foreign investment in the clothing industry	Ongoing reference to literature review prepared during phase 2 Identification of possible policies to propose remedies on the gaps identified during analysis of subsidiaries according to the framework Identification of areas of further research that can serve policy makers to upgrade the clothing industry

(continued)

Table 5.3 (continued)

Phase 1	Phase 2	Phase 3	Phase 4	Phase 5
Preliminary research	Preparation of the framework	Fieldwork	Analysis	Conclusions and policy recommendations
Specification of criteria to select the four clothing manufacturing subsidiaries to be investigated Preparation of the initial questionnaire for the case studies to discuss with clothing industry representative of the Chamber of Façon in Albania Preparation of the first set of interview questions to discuss with relevant institutions and organizations	Finalization of the questionnaire to analyze the four clothing manufacturing subsidiaries Preparation of a tentative schedule of meetings with stakeholders to be followed during the fieldwork Identification of gaps in the data available from authorized institutions	Participation and observations in each department of subsidiaries to understand steps in the production process Meetings and interviews with representatives of associations of the clothing industry and trade unions in Albania Meetings and interviews with representatives of various donors related to the clothing industry in Albania	Discussion of case study analysis with representatives of the Chamber of Façon in Albania	

Source Author's design based on fieldwork

Table 5.4   Criteria for trustworthiness in the case study methodology

No.	Criteria for trustworthiness	Usefulness
1.	Credibility	The steps followed in the case study methodology are documented based on the questionnaire prepared, on the information received through emails, on-site interviews, and when allowed pictures taken in the clothing manufacturing subsidiaries
2.	Transferability	The case study methodology is based on the two pillars of the framework, which can be applied in analyzing subsidiaries in any host territory and in different industries
3.	Dependability	Documentation is assured from the information obtained in written from clothing manufacturing subsidiaries part of this research
4.	Confirmability	In each phase of case study methodology the researcher continuously consulted the president of the Chamber of Façon in Albania, who has more than twenty years of experience and knowledge of the clothing industry in Albania, in the region, and Europe. The researcher obtained additional feedback from the textile department in the Polytechnic University of Tirana and economic experts from the University of Tirana, Faculty of Economics
5.	Prolonged engagement	To assure that the researcher was involved long enough in order to enable the researcher to understand better the characteristics and peculiarities of the phenomena under consideration
6.	Inclusion of primary source material	Thirdly, the chain of evidence was established by collecting detailed information on operations of subsidiaries including production activity, organizational structures, customers, raw materials, suppliers, subcontractors, employment, training of employees, etc.
7.	Purposeful sampling	In order to assure purposeful sampling criteria for sample selection were set based on the traits of the clothing industry in Albania

Source Author's design based on fieldwork

Table 5.5 Validity in case study methodology

No.	Type	Description
1.	Internal validity	Identification of many variables with reference to production, investments, employment, training, sales, clients, types of products of the subsidiary, etc. for each case study Establishment of links between the pillars of the framework during case analysis Connection between variables used to present the clothing industry and foreign investment in Albania with those included in the four cases
2.	External validity	Construction of the framework based on theory Continuous reference to theory while analyzing the four clothing manufacturing subsidiaries Findings, conclusions, and policy recommendations are based on theory introduced in the literature review

Source Author's design based on fieldwork

identified as per the proposed framework and incorporated in policy recommendations. In addition, the variables identified for each case were useful for establishing any relationships with the clothing industry in Albania.

External validity

It is obtained by referring to theory. The framework on which case studies are analyzed is strongly based on extensive literature review. Reference to theory is made throughout the research and especially in drafting the analysis and in drawing conclusions on the four cases. In addition, as presented in Table 5.5 the findings and the conclusions of this research are based on the strains of theory introduced in the literature review chapter. Most importantly, external validity in this research relies on the replication of steps followed in the four clothing manufacturing enterprises were useful not only to observe and understand the dynamics in each case but also to draw generalizations from this research.

5.5   Framework for Case Study Analysis in the Clothing Industry in Albania

The proposed framework in Fig. 5.4 is based on the literature review previously presented. The framework serves as the reference point for analyzing the qualitative effects of foreign clothing manufacturing enterprises in Albania. The proposed framework has two pillars: (i) knowledge transfer to the host territory and (ii) the evolution in the quality of the subsidiary overtime.

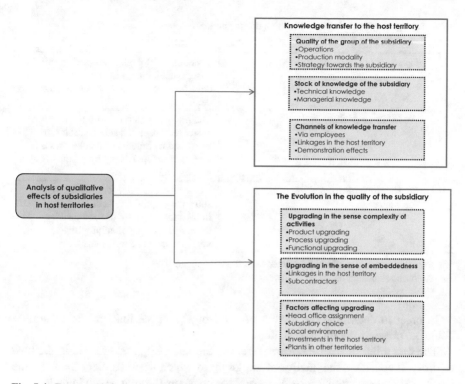

Fig. 5.4 Framework for case study analysis. *Source* Author's design based on fieldwork

Qualitative effects of FDI in the host territory will depend on the quality of the group to which the subsidiary belongs, the quality of the subsidiary, and the channels of knowledge transfer with the agents of the host territory (see Fig. 5.5).

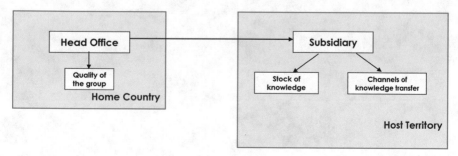

Fig. 5.5 The connection between the group, the subsidiary, and the host territory. *Source* Author's design based on fieldwork

Table 5.6 Quality[a] of the group to which the subsidiary belongs

No.	Quality of the group the subsidiary belongs	Indicator
1.	Operations in the head office	Functions with high gains performed: design, marketing, research and development, purchasing of raw materials, etc.
2.	Production modality	The group depends only on the orders of customers (is only a subcontractor to international brands) or it possesses a degree of independency by producing also its own brand
3.	Strategy towards the activity of the subsidiary	Delegation of complex activities to the subsidiary, appointment of local employees in high management of the subsidiary, training programs of employees, investments made in the host territory

[a]Quality as a potential emitter of knowledge in the host territory
Source Author's design based on fieldwork

In order to determine the quality of the group[1] and the links to the subsidiary Table 5.6 presents the main indicators.

The first pillar of the framework refers to the knowledge transferred in the host territory. It considers the quality of the group to which the subsidiary belongs, the stock of knowledge of the subsidiary, and the channels through which knowledge is transferred in the host territory. Indicators for this pillar are included in Table 5.7.

The second pillar of the framework refers to the evolution in the quality of the subsidiary. This pillar includes: (i) upgrading in the sense of complexity of activities, (ii) upgrading in the sense of embeddedness and (iii) factors that affect upgrading. Indicators of subsidiary upgrading included in Table 5.8.

---

[1]The group will refer to the head office, subsidiaries, subcontractors, form of cooperation with suppliers and customer in each clothing manufacturing enterprise.

Table 5.7 Indicators of knowledge transferred in the host territory

No.	Stock of knowledge of the subsidiary	Areas to consider for clothing manufacturing subsidiaries in Albania
1.	Technical	Quality testing, assembly operations, functions of departments, expansion in the kinds of operations and processes offered over the years, etc.
2.	Managerial	Level of local staff engagement as managers in subsidiaries, responsibilities in selection of employees, financial management, risk diversification, market research, etc.
No.	Channels of technology transfer	Indicators for the research
1.	Via employees	Level of skills in local employees; Position of local employees within the hierarchy of the subsidiary (shop-floor—middle management—top management); rotation within the production process (sewing—cutting—quality testing—supervisor) Training of employees (duration, initial, workshop, technical, fields of knowledge, place of training) Presence of foreign experts, level of interaction with foreign experts, length of visits in the subsidiaries, etc.
2.	Linkages in the host territory	Type of linkages (dependency of sales local supplier/subcontractor to buyer-subsidiary, kind of intermediate product (complex, simple) sold by supplier/subcontractor to subsidiary, price paid to subcontractor/supplier) Counselling of local suppliers/subcontractors by subsidiary
3.	Demonstration effects	Creation of new firms, spin-offs (by ex-employees of subsidiary) Ex-subcontractor of the subsidiary as an independent enterprise producing directly for the final buyer (customer)

Source Author's design based on fieldwork

Table 5.8 Indicators of subsidiary upgrading

No.	Upgrading in the sense of complexity of activities	Indicators for the research
1.	Process upgrading	Evolution of processes through time, moving from assembly to full package, complexity of processes realized since settling in the host territory
2.	Product upgrading	Production of standardized and differentiated products. Changes in the category of finished goods manufactured through years
3.	Functional upgrading	Functions undertaken within the subsidiaries ranging from marketing, design, branding, selection and monitoring of suppliers, etc. Involvement of subsidiaries in functions realized in the head office
No.	Upgrading in the sense of embeddedness	Indicators for the research
1.	Linkages in the host territory	Forward and backward linkages in the host territory, the nature of linkages
2.	Subcontractors	The nature and degree of cooperation with local subcontractors, the number of subcontractors, the functions realized in subcontractors, the level of production, etc.
No.	Factors affecting upgrading at a firm level	Indicators for the research
1.	Head office assignment	Decision of the head office on the responsibilities assigned to subsidiaries in the host territory. Changes and the nature of assigned responsibilities. Degree of autonomy of the subsidiary (investment, selection of suppliers/subcontractors, choosing the employees (workers managers), finding customers, etc.
2.	Subsidiary choice	Decisions of the managers of the subsidiary in the activities in the host territory including establishment of forward and backward linkages. The degree of "freedom" local managers have in operational activities of the subsidiary, their role in selection of suppliers, assignment of responsibilities to subcontractors, etc.
3.	Local environment factors	Identification of any political, social, economic occurrences in the host territory that may have affected operations in case studies

(continued)

Table 5.8   (continued)

No.	Factors affecting upgrading at a firm level	Indicators for the research
4.	Investments in the host territory	Investments in specialized equipment that cannot be easily transferred or used in another industry. High fixed costs associated with the investment made the degree of investment in operational facilities, machines, and ICT of cases in Albania, etc. Expansion in the activity/investment of existing subsidiaries potential for future investments; opening of new subsidiaries in Albania The point in time investments are made in the subsidiary. The more distant in time the investment higher possibilities for moving out. Identification of when the investments were made and how they are related to the activities of the cases
5.	Plants in other territories	Existence of other operational plans of the enterprise and the level of subcontracting in other countries, the importance of other plants or subcontractors in production. Capacity of existing plants, market in which plants sell their products

Source Author's design based on fieldwork

References

1. Baxter P, Jack S (2008) Qualitative case study methodology: study design and implementation for novice researchers. Qual Rep 13(4):544–559
2. Behrman JN, Wallender HW (1976) Transfer of manufacturing technology within multinational enterprises. Ballinger Publishing Company, Cambridge, Massachusetts
3. Bian Y, Xie J, Yang Y, Hao M (2019) Local embeddedness, corporate social capital and Chinese enterprises: the case of Shaanxi FDI firms. Chin Manag Stud 13(4):860–876. https://doi.org/10.1108/CMS-08-2018-0644
4. Blomstrom M, Kokko A (1998) Multinational corporations and spillovers. J Econ Surv 12(2):1–31
5. Boblin SL, Ireland S, Kirkpatrick H, Robertson K (2013) Using stake's qualitative case study approach to explore implementation of evidence-based practice. Qual Health Res 23(9):1267–1275. https://doi.org/10.1177/1049732313502128
6. Callychurn DS, Soobhug K, Hurreeram DK (2014) Key success factors of the apparel manufacturing industry: a case study at company X. In: Proceedings of the world congress of engineering, vol II. Newswood limited, London
7. Cheewatrakoolpong K, Boonprakaikawe J (2015) Factors influencing outward FDI: a case study of Thailand in comparison with Singapore and Malaysia Southeast. Asian J Econ 3(2):123–141
8. Creswell JW (2014) Research design: qualitative, quantitative and mixed methods approaches, 4th edn. Thousand Oaks, Sage, California
9. Denzin NK, Lincoln YS (2000) Handbook of qualitative research, 2nd edn. Sage Publications Inc, London, United Kingdom
10. Dikkaya M, Keles I (2006) A case study of foreign direct investment in Kyrgystan. Central Asian Surv 25(1):146–156

11. Eisenhardt K (1989) Building theories from case study research. Acad Manag Rev 14(4):532–550
12. Evan P (1979) Dependent development. The Alliance of multinational, state and local capital in Brazil. Princeton University Press, Princeton, New Jersey, United States
13. Flyvbjerg B (2006) Five misunderstanding about case-study research. Qual Inq 12(2):219–245
14. Gerring J (2004) What is a case study and what is it good for? Am Polit Sci Rev 98(2):341–354
15. Gioia DA, Corley KG, Hamilton AL (2013) Seeking qualitative Rigor in inductive research: notes on the Gioia methodology. Organ Res Methods 16(1):15–31. https://doi.org/10.1177/1094428112452151
16. Golafshani N (2003) Understanding reliability and validity in qualitative research. Qual Rep 8(4):597–607
17. Guba EG, Lincoln YS (1985) Naturalistic inquiry. Sage Publications, Newbury Park, California
18. Helper S (2000) Economists and field research: you can observe a lot just by watching. Am Econ Rev 90(2):228–232
19. Johanson J, Widersheim-Paul F (1975) The internalization of the firm-four swedish case studies. J Manag Stud 305–322
20. Johansson A, Fristedt S, Boström M, Björklund A, Wagman P (2019) Occupational challenges and adaptations of vulnerable EU citizens from Romania begging in Sweden. J Occup Sci 26(2):200–210
21. Kacani J (2016) Towards knowledge based flexibility for manufacturing enterprises: with a case study. Int J Intell Enterp 4(3):204–226
22. Killam L (2013) Research terminology simplified: paradigms, ontology, epistemology and methodology. Sudbury
23. Langdon S (1981) Multinational corporations in the political economy of Kenya. Macmillan, London, United Kingdom
24. Larrain FB, Lopez- Calva LF, Rodriguez-Clare A (2000) Intel: a case study of foreign direct investment in Central America. Working Paper No. 58, Center for International Development at Harvard University, Cambridge, Massachusetts
25. Merriam SB (2009) Qualitative research: a guide to design and implementation, 3rd edn. Jossey-Bass, California
26. Miles R (2015) Complexity, representation and practice: case study as method and methodology. Issues Educ Res 25(3):309–318
27. Miles MB, Huberman AM, Saldaña J (2014) Qualitative data analysis: a methods sourcebook and the coding manual for qualitative researchers. Thousand Oaks, Sage Publications, California
28. Munday M, Morris J, Wilkinson B (1995) Factories or warehouses? A Welsh perspective on Japanese transplant manufacturing. Reg Stud 29(1):1–17
29. Nayak AKJR (2005) FDI model in emerging economies: case of Suzuki Motor corporation in India. J Am Acad Bus 238–246
30. Oppenheim AN (2000) Questionnaire, design, interviewing and attitude measurement. Bloomsbury Academic, United Kingdom
31. Osano HM, Koine PW (2016) Role of foreign direct investment on technology transfer and economic growth in Kenya: a case of the energy sector. J Innov Entrepreneurship 5:31. https://doi.org/10.1186/s13731-016-0059-3
32. Paprzycki R (2006) The impact of foreign direct investment in Japan: case studies of the automobile, finance, and health care industries. Discussion Series Papers, No. 141. Institute of Economic Research, Hitotsubashi University, Tokyo
33. Patton MQ (1990) Qualitative evaluation and research methods, 2nd edn. Sage Publications, Newbury Park, California
34. Phelps NA, Lovering J, Morgan K (1998) Tying the firm to the region or tying the region to the firm? Early observations on the case of LG in South wales. Eur Urban Reg Stud 5(5):119–137
35. Pirsig RM (1989) Zen and the art of motorcycle maintenance. A Corgi Books Publication, United States
36. Platt J (1992) Case study in American methodological thought. Curr Sociol 40:17–48

37. Ridder HG (2017) The theory contribution of case study research designs. Bus Res 10(2):281–305
38. Schoenberger E (1991) The Corporate interview as a research method in economic geography. Prof Geogr 43(2):180–189
39. Stake RE (1995) The art of case study research. Sage Publications Inc, New York, United States
40. Stake (2005) Multiple case studies analysis. The Guilford Press, New York
41. Suddaby R, Hardy C, Huy (2017) Introduction to special topic forum: where are the new theories of organization? Acad Manag Rev 36(2):236–246
42. Thomas G (2010) Doing case study: abduction not induction, phronesis not theory. Qual Inq 16(7):575–582
43. Tsang EWK (2014) Generalizing from research findings: the merits of case studies. Int J Manag Rev 16(4):369–383
44. Van Maanen J (1988) Tales of the field: on writing ethnography. Chicago Guides to Writing, Editing, and Publishing
45. Wadham H, Warren RC (2014) Telling organizational tales: the extended case method in practice. Organ Res Methods 17(1):5–22. https://doi.org/10.1177/1094428113513619
46. Wen Y, Shih C, Agrafiotes K, Sinha P (2014) New product development by a textile and apparel manufacturer: a case study from Taiwan. J Text Inst 105(9):905–919
47. Yin RK (2003) Case study methodology, 3rd edn. Sage Publications Inc., New York

Chapter 6
The Evolution of Clothing Manufacturing Subsidiaries

Abstract This chapter presents the case studies that consists in four clothing manufacturing subsidiaries that started production activity in emerging economies in the early 1990s. The operational activity of the head office and the production activity of subsidiaries in the host territory is presented based on the intensive fieldwork undertaken within production facilities. Presentation of case studies includes the organizational structure, production process flow together with the capacity building of local employees and linkages with suppliers. With regard to the value added activities global value chain in the clothing industry a comparison is made between the head office and the clothing manufacturing subsidiaries.

Keywords Case studies · Process flow · Capacity building · Value adding activities

6.1 Overview of Clothing Manufacturing Enterprises in Albania

This chapter introduces the four case studies subject to this research (see Table 6.1). Presentation of each case starts with the introduction of the main characteristics of the region in which the subsidiary is located, it continues with a description of the group to which the subsidiary belongs and with an in-depth presentation of the subsidiary including its history, production activity, investments, organizational structure, staff training, and compensation. The group to which the four cases belong are foreign owned and they exercise their operational activity mostly in two countries.

6.2 Shqiperia Trikot sh.p.k

This section presents the first case study of the research. It is dedicated to the operational and production activity of the Italian enterprise Cotonella S.p.A and its subsidiary Shqiperia Trikot sh.p.k. Presentation of the case starts with the activity of the group and continues with the activity of the subsidiary and its subcontractors.

© Springer Nature Switzerland AG 2020   163
J. Kacani, *A Data-Centric Approach to Breaking the FDI Trap Through Integration
in Global Value Chains*, Lecture Notes on Data Engineering and Communications
Technologies 50, https://doi.org/10.1007/978-3-030-43189-1_6

Table 6.1 Overview of clothing manufacturing subsidiaries

Indicator	Shqiperia Trikot sh.p.k	Naber Konfeksion sh.p.k	Valcuvia Alba sh.p.k	Industria Ballkanike sh.p.k
Realization of the field work	February–April, 2016	January–March, 2016	May–June, 2016	April–June, 2016
Head office/group to which the clothing manufacturing subsidiary belongs	Cotonella S.p.A, an Italian enterprise that manufactures intimate apparel for its own brand, for retailers, and branded marketers	Naber Moden, a German enterprise that manufactures apparel for its own brand, for retailers, and branded marketers	Valcuvia Alba s.r.l an Italian enterprise that manufactures intimate apparel for retailers and branded marketers	Industria Ballkanike, a Greek enterprise that manufactures apparel for retailers and branded marketers
Categorization of production modalities	Full package with brand	Full package with brand	Full package	Cut, make, trim
Kinds of clothing articles	Underwear for men, women, and children, swimwear, night gowns, pyjamas	Trousers, skirts, parkas, blouses, jeans, coats, fur coats, traditional dresses	Underwear for men, women, children swimwear, nightgowns, pyjamas	Sport outfits, shirts, shorts, swimwear
Degree of competition in the product market	Medium competition in the market for its own brand products Strong competition in the market for assembly of other brands	Medium competition in the market for its own brand products Strong competition in the market for assembly of other brands	Strong competition in the market for assembly of other brands	Strong competition in the market for assembly of other brands
Principal motivation for investment in the host territory	Reduction in the delivery time of finished articles, cheap labor force, and social cultural similarities found in Albania	Reduction in the delivery time of finished article, cheap labor force, proximity of the country to the EU market	Cheap labor force and the skills of the labor force in clothing manufacturing	Cheap labor force and the skills of the labor force in clothing manufacturing
Date of start of production in Albania	1995	1995	1992	1993
Employment in the subsidiary	78	850	800	529

(continued)

Table 6.1 (continued)

Indicator	Shqiperia Trikot sh.p.k	Naber Konfeksion sh.p.k	Valcuvia Alba sh.p.k	Industria Ballkanike sh.p.k
Operations performed in the subsidiary	Assembly, cutting, comprehensive quality control of raw materials, packaging, distribution of finished products, research and development in neighboring countries of the Balkan region, brand marketing, distribution of finished products	Assembly, cutting, quality control of raw materials, preparation of models for cutting, buttoning, embroidery, labeling, ironing, packaging, distribution of finished products	Limited quality control of raw materials, cutting, assembly, packaging, distribution of finished products	Limited quality control of raw materials, cutting, assembly, printing, embroidery, stamping, pressing, packaging, distribution of finished products
Investments made in the host territory	Ownership of production facilities (land and buildings), information technology (system and equipment), quality control laboratories, variety of machines (printing, pressing, ultrasound, cutting, sewing)	Ownership of production facilities (land and buildings), information technology (system and equipment), quality control laboratories, cutting machines, sewing machines, embroidery machines, vehicles, power supply generators	Ownership of production facilities (land and buildings),information technology (system and equipment), quality control laboratories, variety of machines (printing, pressing, ultrasound, cutting, sewing)	Ownership of production facilities (land and buildings), cutting machines, sewing machines, embroidery machines, information technology (system & equipment), vehicles, power supply generators
% of total production of the head office manufactured in Albania	98	95	75	100
Modality	Full package with design	Full package with design	Full package with design	Full package

(continued)

Table 6.1 (continued)

Indicator	Shqiperia Trikot sh.p.k	Naber Konfeksion sh.p.k	Valcuvia Alba sh.p.k	Industria Ballkanike sh.p.k
Relations with local suppliers of raw materials and intermediary products	Totally absent	Totally absent	Totally absent	Totally absent
High and middle management in the subsidiary	Occupied by local staff only	Occupied by expatriates and local staff	Occupied by expatriates and local staff	Occupied by expatriates and local staff
Possession of subsidiaries in other countries	No	No	No	No
Possession of subcontractors in other countries	China and India	FYR Macedonia	Serbia and North Africa	No
The region in which the subsidiary is located	Shkodra	Durres	Berat	Korce
GDP per capita of the region (EUR) in 2017	2,902	3,994	3,364	3,001
GDP of the region as % GDP of Albania in 2017	5.17	9.91	3.77	5.51
Foreign owned enterprises compared to the total number of enterprises in the region (%) in 2017	2.64	11.6	0.05	1.64

Source Shqiperia Trikot sh.p.k, Naber Konfeksion sh.p.k. Valcuvia Alba sh.p.k, Industria Ballkanike sh.p.k, INSTAT

6.2.1 Overview of the Region of Shkodra

The region of Shkodra is situated in the north of Albania sharing borders with Montenegro (see Fig. 6.1). It is one of the oldest and most historic places in Albania,

Fig. 6.1 The map of Albania with the region of Shkodra circled. *Source* www.europe-atlas.com

Table 6.2 Table presenting the region of Shkodra (2017)

Indicator (2017)	Region of Shkodra	Albania
Population	207,924	2,876,591
Total GDP in (Mln/EUR)	599	11,564
GDP per capita (EUR)	2902	4024
GDP composition in (%) of the total GDP		
Agriculture, forestry, and fishery (%)	30.80	22.70
Industry (%)	16.60	13.90
Construction (%)	8.30	10.20
Trade, transport, hotels, and restaurants (%)	14.60	18.10
Information and communication (%)	3.20	3.50
Financial and insurance activities (%)	2.10	2.80
Real estate activities (%)	6.20	6.60
Scientific, professional, administrative, and supporting activities (%)	3.40	6.70
Public administration, education, health, and social activities (%)	12.80	12.50
Artistic, entertaining, and related activities (%)	2.00	3.00
Number of enterprises	11,309	162,452
Number of foreign owned enterprises	98	3704

Source INSTAT and the Chamber of Façon in Albania

as well as an important cultural and economic center. The region of Shkodra had a population of 207,924 inhabitants in 2017[1]. In this region can be found the largest hydropower plants of Albania. Recently, the region has attracted major Austrian companies that have developed two additional hydropower plants in 2012.

In 2017, the GDP in current prices for the region of Shkodra amounted to 599 million EUR contributing to the Albanian GDP with 5.17% (see Table 6.2). The GDP per capita for the region in 2017 was 2,902 EUR. The main economic activity of the region of Shkodra is primarily based, which contributes by 30.80% in the region's GDP. This region has also a well-developed fish processing industry, which is based on the variety of fish and related species that are found in the Adriatic Sea.

Trade (wholesale and retail) and tourism related economic activities (restaurants, hotels, entertainment, etc.) have a share of 14.6% in the region's GDP. The main trade partners of the region are Montenegro, Croatia, and Serbia.

After the establishment of the market economy in Albania, several foreign clothing manufacturing enterprises started to operate in the region of Shkodra, mainly originating from Italy. Handcrafting (embroidery and carpet making, etc.) is an area of the regional economy, which is flourishing due to the high demand from foreign

---

[1]In Albania with the few exceptions regions are named after the largest city located within the geographical area of the region.

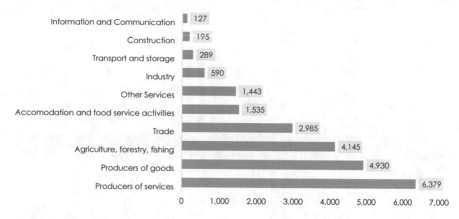

Fig. 6.2 Number of enterprises in the region of Shkodra (2018). *Source* National Statistical Institute (INSTAT), Business Register 2018

visitors. Shkodra is well-known for manufacturing of luxurious Venetian masks. Enterprises that engage in the manufacturing of masks are major suppliers for the carnivals organized annually in Venice.

In 2018, the region of Shkodra had 11,309 active enterprises, 6.94% of the total active enterprises in the country (see Fig. 6.2). Of the active enterprises, females own 2581 while 104 are foreign owned while 70 enterprises are jointly owned between foreigner and local citizens. The region is dominated by SMEs amounting to 10,472 that employ up to four workers. Enterprises employing up to 49 persons are 374 while those having over 50 employees are only 98.

6.2.2 The Cotonella S.p.A Group

This section presents the history and the operating activity of Cotonella S.p.A. This section includes the history, the customers, the output, customers, suppliers of raw materials, logistics the organizational structure, and value added activities performed by Cotonella S.p.A.

The group is composed of offices in Italy and a production unit in the form of a subsidiary in Albania. The head office and the design department are located in Sonico (Brescia) while logistics unit is located in Malonno (Brescia). Cotonella S.p.A has also offices in Val di Chiana (Tuscany), Franciacorta (Brescia) and Milan. In 2015, Cotonella S.p.A had 90 employees. In Albania, the group has its subsidiary Shqiperia Trikot sh.p.k located in the region of Shkodra.

In 1972, Cotonella S.p.A started its operations as a small clothing manufacturing enterprise for the production of intimate apparel in Edolo located in the province of Brescia in north of Italy (see Fig. 6.3).

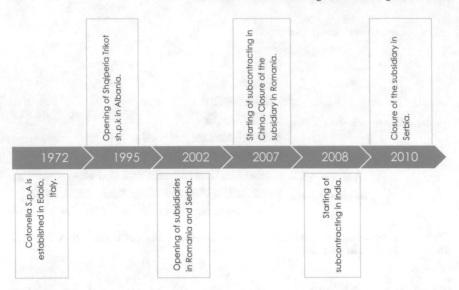

Fig. 6.3 Main events in the history of Cotonella S.p.A. *Source* Based on the information provided by Shqiperia Trikot sh.p.k

In 1995, it opened in Albania its subsidiary Shqiperia Trikot sh.p.k. In 2000s the activity expanded both with opening of additional subsidiaries in Romania and Serbia and with subcontracting in China and India. By 2010, the two subsidiaries in Romania and Serbia closed down limiting the activity of Cotonella S.p.A to production in Albania and subcontracting in Asia.

Cotonella S.p.A, is able to manufacture through its subsidiary in Albania and subcontractors in Asia all kinds of intimate apparel for men, women, boys, girls, and babies both for its own brand and other customers. The main product lines for its own brand are shown in Table 6.3. The daily production capacity for all goods (underwear, loungewear, and nightwear) is about 50,000 items.

In addition to its own brand, Cotonella S.p.A has manufactured intimate apparel also for international Italian and European brand such as Coop, Esselunga, Magnun, Galeria Kaufhof, Guess, and Calvin Klein. The annual output of Cotonella S.p.A exceeds that of 16 million pieces per year. Over 98% of the total output including goods for "Cotonella" brand and orders from customers are manufactured by its subsidiary Shqiperia Trikot sh.p.k located in the region of Shkodra.

Cotonella S.p.A is lead by its president that is the highest decision taking authority of the enterprise. Since 2011, he is the president of the Italian-Albanian Chamber of Commerce. In 2007, revenues of Cotonella S.p.A reached 31 million EUR while in 2008 due to the global financial crisis Cotonella S.p.A suffered a loss of 2 million EUR. In 2009 and in 2010, the enterprise started to recover and its revenue collection was between 28 and 30 million EUR.

On this experience president Zannier says "I am pleased about the recovery. We have already developed new projects that are about to start. In the short run, we

Table 6.3 The product lines of Cotonella brand

No.	Lines of Cotonella brand	Description of the line
1.	Premier	Line focused on articles with details in design
2.	Original	Traditional underwear
3.	Optima	Underwear for sale in commercial centers
4.	Invisible	Underwear line that is not visible from cloths
5.	Silhouette	Underwear line focusing in the body anatomy
6.	In and out	Models visible out of sweaters and jackets
7.	Boy and girl	Underwear line for children
8.	Cotonella come Seta	Intimate underwear of fine fabric
9.	L'Altra Cotonella	Intimate underwear at convenient prices
10.	Night and day	A line of nightgowns and pyjamas
11.	Natale	Intimate underwear designed for end year holidays

Source Based on the information provided by Cotonella S.p.A

are thinking about new products and innovative distribution systems. Making new investments is the only way to respond to difficult times".[2]

Previously, Cotonella S.p.A had subsidiaries in Rumania and Serbia (see Table 6.4). These subsidiaries offered only sewing and packaging services. Subsidiaries cooperated also with five subcontractors Romania and three in Serbia.

Cotonella S.p.A decided to terminate its activity in Romania when the labor cost started to go up after it joined the European Union and to exist Serbia in 2010 when the new production site in Albania became operational. However, Cotonella S.p.A still has subcontractors in China and India that produce mainly corsetry and pajamas (see Table 6.5).

In addition to Cotonella brand that covers 50% of total production, Cotonella S.p.A produces intimate apparel for a number of customers through its subsidiary in Albania (among others retailers like Coop, Esselunga, and Avon). For each customer

Table 6.4 Former subsidiaries of Cotonella S.p.A in Rumania and Serbia

No.	Indicator	Subsidiary in Rumania	Subsidiary in Serbia
1.	Starting date of the activity	2002	2002
2.	Articles	Slip for men and women, bras, and shirts	Slip for men and women, bras
3.	Activities in production cycle	Sewing and packaging	Sewing and packaging
4.	Number of subcontractors	5	3
5.	Exit date of the country	2007	2010

Source Based on the information provided by Cotonella S.p.A

---

[2]This statement was made during an interview in one of the visits in Shqiperia Trikot sh.p.k.

Table 6.5 Subcontractors of Cotonella S.p.A in China and India

No.	Indicator	China	India
1.	Starting date of the production activity	2007	2008
2.	Types of items produced	Corsetry, pajamas, slip	Corsetry, shirts
3.	The average quantity[a] of articles produced annually	100,200	47,730
4.	% in the annual total output of Cotonella S.p.A in 2015	1.5	0.5

[a]The average quantity of articles manufactured annually is calculated by taking the average of the quantity manufactured (2010–2015)
Source Based on the information provided by Cotonella S.p.A

it is able to produce a variety of articles like in case of Coop and Gerko when the number goes to 30 (see Table 6.6).

In the 1990s, Cotonella S.p.A was responsible for the selection of suppliers and imports of raw materials and accessories (slingshot, thread, labels) which are sent to its subsidiary Shqiperia Trikot sh.p.k. Only after the 2000s Cotonella S.p.A made Shqiperia Trikot sh.p.k responsible for the import of fabrics and accessories required for sewing and packing, while it remains responsible for the selection of suppliers. The subsidiary imports raw materials mainly from Turkey and a minor fraction from Italy, Romania, Bulgaria, and China. For an output of 12–16 million pieces a year, it requires around 500–700 tons of imported fabric. In recent years, imports from Turkey have increased considerably as the result of the agreement between the customs office of Albania and Turkey, which consists in preferential rates on these imports.

Table 6.6 Main customers of Cotonella S.p.A

No.	Main customers	Category	Number of various articles	Output in % for 2014
1.	Coop	Slip and shirts for men, women, and children	30	15
2.	Esselunga	Slip for men and women	20	10
3.	Gerko	Slip and shirts for men, women, and children	30	10
4.	Avon	Slip for women	25	5
5.	Giannini distribution	Slip and shirts for men, women, and children	30	5
6.	Kaufhof	Slip for women	6	5

Source Based on the information provided by Shqiperia Trikot sh.p.k

Raw materials are imported based on price-quality ratio and according to the preferences of the customers of Cotonella S.p.A. The quality of raw materials is guaranteed in Shqiperia Trikot sh.p.k through continuous tests and controls immediately after they are stored in the premises of the subsidiary. In case non-conformities are identified during quality control, suppliers are immediately notified for their presence. In case suppliers are skeptical on testing procedures, they are invited to observe testing procedures and results in the laboratories of Shqiperia Trikot sh.p.k. The rejection rate of raw materials has been small and sporadic. In addition, Cotonella S.p.A schedules regular meetings every three months during which executives and technicians of Shqiperia Trikot sh.p.k coming from the head office: (i) get familiar with the newly manufactured fabrics, (ii) make recommendations on improving the technical specifications of existing fabrics, and (iii) place orders on specific fabrics asked from customers that are not available in the market. Local suppliers are absent in operations of Cotonella S.p.A. Up to now no local firms are able to produce raw materials able to meet the stringent requirements set by Cotonella S.p.A.

Cotonella S.p.A acts both as a distributer of finished articles to end customers and as a supervisor of the logistic network created through its subsidiary. It has assigned Shqiperia Trikot sh.p.k to deliver finished goods to end customers. The subsidiary delivers the goods of the "Cotonella" brand across the selling points and customers assigned by the head office. The head office has financed the purchase of large vehicles. These vehicles use the European road transportation system to assign finished articles according to the instructions of the logistic department in the head office. For selling points that are not located in nearby countries and that are not convenient and lengthy, to be reached by the road transportation system Shqiperia Trikot sh.p.k delivers finished articles in the store houses of Cotonella S.p.A. It is the responsibility of the logistic department within Cotonella S.p.A to assign respective goods to selling points.

Key departments are located in the head office of Cotonella S.p.A. These departments closely cooperate with Shqiperia Trikot sh.p.k (see Fig. 6.4).

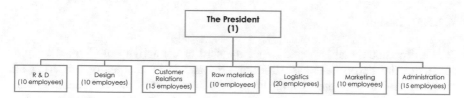

Fig. 6.4 Organizational structure of Cotonella S.p.A. *Source* Author's drawing based on the information provided by Cotonella S.p.A

6.2.2.1 Research and Development (R&D)

It is a department with 10 employees with one of them serving as the department head. The main objective of the department is to analyze the global developments of the clothing industry and to identify the techniques to improve the quality of goods of the "Cotonella" brand. The department conducts research on: (i) manufacturing of new intimate apparel, (ii) creating innovative products that allow for maximal customization to the human body anatomy, and (iii) generating new goods in accordance with the changes in the preferences of customers.

6.2.2.2 Raw Materials

It is a department with 10 employees that deals mainly with orders and control of raw materials, especially for fabrics used in the manufacturing of intimate apparel. In the raw materials department is placed the internal quality control laboratory. This laboratory is equipped with modern instruments able to perform quality testing for all raw materials. Raw materials used in manufacturing of intimate apparel are certified according to OEKO-TEX 100 standard, which guarantees the highest levels of hygiene in accordance with environmental regulations indicating the absence of potentially harmful substances. In 2001, the quality system managed to get the UNI EN ISO 9001:2008 certification ensuring the quality of the enterprise to carry out constant checks to obtain finished articles acceptable to international standards.

All intimate apparel is equipped with an antibacterial silver ionized AB100 fortress, which prevents the formation of bacteria that cause irritation. This extra special antibacterial fortress is placed on all lines of production of the "Cotonella" brand. For orders other than Cotonella brand, intimate apparel is manufactured according to their specifications set by clients of Cotonella S.p.A who are well integrated into global production chains.

6.2.2.3 Design

It is a department that is present since the start of the activity of Cotonella S.p.A in 1972. The department has 10 employees and is in charge for preparing all styles and models for the 11 intimate apparel lines of "Cotonella" brand. Furthermore, this department adapts the designs received from various customers into the manufacturing process of Cotonella S.p.A.

6.2.2.4 Marketing

It is a department that deals with sales of "Cotonella" brand and in promoting the quality of services to potential international customers. It has 10 employees. This department organizes the advertising campaign of "Cotonella" brand and identifies new selling points for its products. It is supported by the logistics department, which ensures that all the selling points have the supply of intimate apparel according to their needs.

6.2.2.5 Customer Relations

It is a department that has 15 employees and is responsible to obtain continuous feedback from customers in order to identify new services to be provided to customers.

6.2.2.6 Logistics Department

It is a department with 25 employees who are responsible for the delivery of finished articles to customers and for the placement of "Cotonella" brand in respective stores and shopping centers.

6.2.2.7 Administration

It is a department in the head office that includes accounting department, human resources, drivers, security, etc. In total, 15 employees are part of this department.

Figure 6.5 presents the value added activities occurring in Cotonella S.p.A in the clothing industry according to value added activities in the global value chain

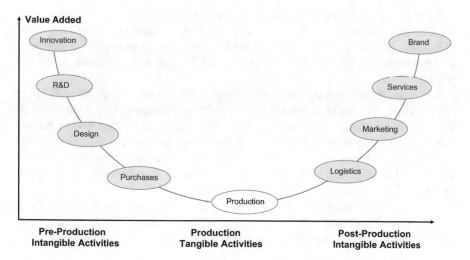

Fig. 6.5 Value adding activities in Cotonella S.p.A. *Source* Based on the information provided by Cotonella S.p.A and Shqiperia Trikot sh.p.k

presented in Chap. 3. The head office is responsible for all activities except production that occurs in the subsidiary. However, the subsidiary is partially involved activities that go beyond production.

6.2.3   The Subsidiary Shqiperia Trikot sh.p.k Clothing Manufacturing Enterprise

This section presents the operational activity of Shqiperia Trikot sh.p.k including organizational structure, production, and investment. Training and compensation.

Shqiperia Trikot sh.p.k was established in 1995. President Zannier says, "in Albania we manufacture the majority of our products, an experience that has been highly satisfactory both at a professional and personal level".[3] Cotonella S.p.A decided to produce 98% of its annual output in Albania because of:

- the favorable political and economic conditions;
- Shqiperia Trikot sh.p.k since its establishment in 1995 has demonstrated a continuous improvement in its performance;
- easy access to the head office of Cotonella S.p.A which can be reached through the international road transportation system that connects the north of Albania with the north of Italy. The short distance to Italy improves logistics and accelerates the delivery of finished articles to every customer;
- ease of operating in the local economy due to social and cultural similarities between the two neighboring countries and of ease of communication as about 90% of the Albanian population speaks the Italian language.

In 1998, after three years of operations Shqiperia Trikot sh.p.k started to work for the first time with two shifts because of its satisfactory performance in manufacturing of intimate apparel (see Fig. 6.6). Within five years, it managed to have a staff level of 450 employees and was able to produce between 10 and 12 million pieces per year.

The large volume of production led to the second restructuring of Shqiperia Trikot sh.p.k, which consisted in outsourcing of production into five main units that operate today as its subcontractors. The main objective of decentralization was to achieve a higher flexibility in production and a better diversification of production risk. In addition, restructuring in production and outsourcing of assembly services occurred as Shqiperia Trikot sh.p.k aims to focus on functions that are more complex and move up in the value chain of the clothing industry.

The process of decentralization proved successful not only for Cotonella S.p.A and Shqiperia Trikot sh.p.k but also for its employees and associates. Employees with the best working performance and the most experienced technicians were given the

---

[3]This statement was made during a short interview with the President in the production facilities of Shqiperia Trikot sh.p.k.

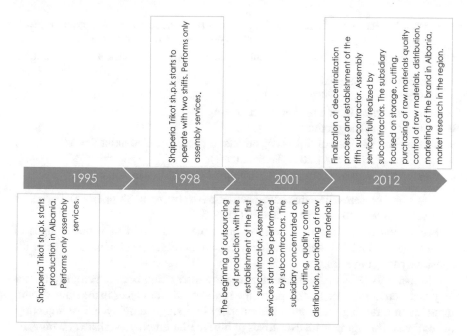

Fig. 6.6 Main events in the history of Shqiperia Trikot sh.p.k. *Source* Based on the information provided by Shqiperia Trikot sh.p.k

opportunity to become in charge for the management of subcontracting firms. Subcontractors are separate fiscal entities and possess the buildings in which assembly services are performed while the machines (sewing, etc.) are provided by Cotonella S.p.A. Cotonella S.p.A and Shqiperia Trikot sh.p.k provide a continuous support to each subcontractor especially in improving manufacturing productivity and in strengthening staff capacities.

After the decentralization process was finalized, Shqiperia Trikot sh.p.k remained with a staff of 78 employees (see Table 6.7) while five subcontractors employ directly more than 400 employees. The new organization of production permits Shqiperia Trikot sh.p.k to provide not only assembly services but also cutting, quality control of raw materials, and delivery of finished articles to end customer.

Table 6.7 Level of employment in Shqiperia Trikot sh.p.k

No.	Year	Level of employment
1.	1995	82
2.	1998	450
3.	2010	120
4.	2015	78

Source Based on the information provided by Shqiperia Trikot sh.p.k

Currently, Shqiperia Trikot sh.p.k has taken the first steps in research and development activities and in marketing of Cotonella brand through the opening of new stores in the Balkan Peninsula. Current services of Shqiperia Trikot sh.p.k include:

- storage of raw materials;
- control of raw materials;
- cutting and preparation of fabrics for sewing;
- control of finished products;
- packing and delivery of finished products to clients of Cotonella S.p.A;
- market research in the Balkan Peninsula;
- management and marketing of "Cotonella" brand stores in Albania.

Figure 6.7 presents the value added activities occurring in Shqiperia Trikot sh.p.k with value added activities presented in Chap. 3.

Production is realized based on the principle of division of various stages of production, which has brought over the years an increased level of production of the subsidiary (see Fig. 6.8).

Shqiperia Trikot sh.p.k aims to increase the interaction between employees in a way that a process and a task are performed by several employees in the same department and not turning them into the monopoly of a single employee. Consequently, the entire staff ranging from executives to blue-collar employees share and transmit knowledge among each other and express their recommendations on particular tasks and processes.

The highest executive of Shqiperia Trikot sh.p.k is the general administrator who coordinates production through the company director and the director for finance administration. These refer also to high management level of the subsidiary.

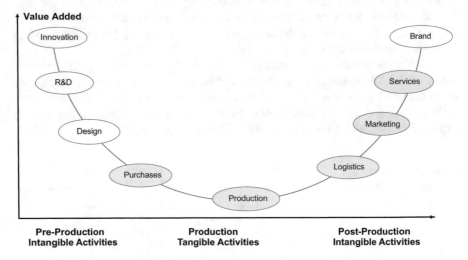

Fig. 6.7 Value adding activities in Shqiperia Trikot sh.p.k. *Source* Based on the information provided by Shqiperia Trikot sh.p.k

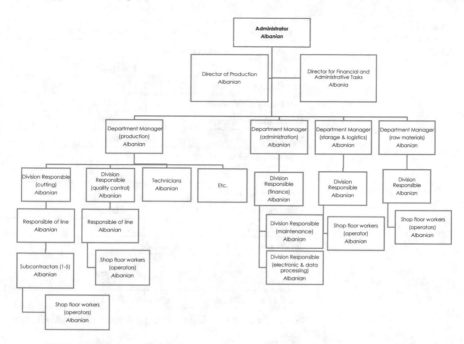

Fig. 6.8 Organizational structure of Shqiperia Trikot sh.p.k. *Source* Author's drawing based on the information provided by Shqiperia Trikot sh.p.k

The company director directly supervises the departments of: (i) storage of raw materials, (ii) cutting department, (iii) storage of finished goods and logistics, and (iv) control of the quality in the testing laboratories. Each of these units has its management structure, which consists of the department manager, the deputy manager of the unit, and responsible for divisions in the department. Department managers are also considered as middle level management of the subsidiary while responsible for divisions are regarded as low-level management.

The department of raw materials was established in 2008. Previously it was the only division within the production department. The purpose of this department is the storage of raw materials including fabrics and accessories and the execution of physical and chemical controls on all types of fabrics used in manufacturing of intimate apparel. The cutting department was created in 2009. Previously it was only a division within the production department. This department is responsible for cutting raw materials, which has been subject to the quality control tests in the department of raw materials. In this department, raw materials are cut into samples and are prepared for sewing. These samples are distributed to subcontractors to continue with sewing and packing processes. To continue, the department of storage of finished articles and logistics was also created in 2008. This department stores all finished goods that are manufactured by subcontractors. Furthermore, this department is responsible for the quality control of finished products delivered to Shqiperia Trikot sh.p.k by its subcontractors and for exporting them directly to customers of Cotonella S.p.A.

Table 6.8 Investments in Shqiperia Trikot sh.p.k (2010–2015)

Category of investment	Amount in EUR
Investments in operating facilities	
Building	2,600,000
Ventilation	300,000
Electricity generators	450,000
Guest house/canteen	200,000
Security	50,000
Investments in machines	
Cutting machines	700,000
Sewing machines	1,000,000
Printing machines	150,000
Quality testing	350,000
Investments in information technology	
Hardware	155,000
Software	45,000
Total of investments	6,000,000

Source Based on the information provided by Shqiperia Trikot sh.p.k

The director for finance administration directly supervises three divisions: (i) finance, (ii) maintenance and (iii) electronic data processing. These divisions are all included in the administration department of the subsidiary. They have a supportive role in the daily operations of Shqiperia Trikot sh.p.k.

Up to now, capital investments focus on: (i) production facilities including storage, administration, and security, (i) machines used in production, and (iii) information technology (see Table 6.8).

In 1995, the initial investment amounted to 190,000 EUR. The highest level of capital investments in production facilities of Shqiperia Trikot sh.p.k was made between 2006 and 2010 during which the new manufacturing premises of 25,000 m² costing 3.2 million EUR was built in the suburbs of the city of Shkodra. Executives and the administration occupy single space offices. The new premises have: (i) a canteen of 56 places for its employees, (ii) fully furnished apartments for professionals and technicians visiting Shqiperia Trikot sh.p.k for a short period of time, (iii) large elevators that permit transportation of high amounts of finished goods, and (iv) a lighting system which is customized depending on the process needed to be completed. Storage spaces in the new facilities have an advanced thermo-isolation system to ensure protection from moisture or extreme heating. In addition, the cutting department is equipped with a ventilation system that absorbs the majority of the waste generated during the cutting of various fabrics. Since 2010, when the new production facilities became operational the total investment has accumulated to 3.8 million EUR.

Another objective of Cotonella S.p.A is to manufacture intimate apparel based on modern machines. The initial investment in machines amounted to 295,000 EUR. In the new production facilities, additional investments were made in sewing machines.

Shqiperia Trikot sh.p.k is the one of the few inward processing companies in Albania that has sewing machines working based on particle filtering technology. This technology minimizes noise while sewing and absorb most of the particles emitted by working with various fabrics. The investment made in sewing machines amounts to 1,000,000 EUR. After the decentralization, the beneficiaries of the investment on machineries are only the subcontractors as in Shqiperia Trikot sh.p.k sewing is no longer available. The technology used in cutting is one of the latest editions of Gerber[4] production and it is almost fully automated. The amount spent on cutting machines is 700,000 EUR or 32% of the total investment. The investment made in machineries that are used to print the designs received from the head office is 150,000 EUR.

Today, the information technology is extended to all administration of Shqiperia Trikot sh.p.k. Investments in information technology are ongoing. The initial investment was 45,000 EUR followed by an additional investment of 155,000 EUR in 2015. The additional investment was mainly dedicated to the distribution system that is supported by high storage capacity servers. The distribution system operates with special software that identifies the location of finished goods until they reach their final destination.

Investments are made also in laboratories used for testing of raw materials and quality control of finished articles. These laboratories able to perform on different types of fabric various physical and chemical tests including: weight, horizontal and vertical flexibility, washing stability, resistance, gentleness, level of acid proof alcalino (loss of color from sweat). Tests for elasticity required in manufacturing of intimate apparel include: mechanical tight, elasticity, color degradation during washing, gentleness, resistance to washing of 1,000 cycles. The testing laboratories have a digital and a modern controlling system that make it possible to eliminate any adverse effect in the health and hygiene of customers. All intimate apparel starting from, shorts, jackets, pajamas, lingerie, to those for babies and children are guaranteed by quality checks that are performed by specialized technicians in the testing laboratory of Shqiperia Trikot sh.p.k. Statistical data are kept at each stage of control. The amount invested in testing and quality control laboratories is 350,000 EUR.

Finally, investments in information technology are also made for the administration of Shqiperia Trikot sh.p.k, which operates with an I core 7 technology. All departments and their units are equipped with computers and video cameras that used by employees to have video conferences with Italian technicians when they encounter difficulties during manufacturing addressing every obstacle that postpones production.

---

[4]Gerber Technology is a world leader in providing sophisticated automated manufacturing systems. The company serves 25,000 customers including more than 100 Fortune 500 companies, in the aerospace, apparel, retail, technical textiles, furniture, and transportation interiors industries in 130 countries.

Table 6.9 Output of Shqiperia Trikot sh.p.k and subcontractors (2010–2015)

No.	Item	Output (pieces) 2010	Output (pieces) 2015	Change in output (%)
1.	Slip	10,000,000	11,800,000	18.00
2.	Shirts	450,000	1,300,000	188.89
3.	Bras	15,000	70,000	366.67
4.	Pajamas	2,000	10,000	400.00
5.	Total	10,467,000	13,180,000	25.92

Source Based on the information provided by Shqiperia Trikot sh.p.k

The strategy of production in Cotonella S.p.A is oriented toward a gradual decline in the level of manufacturing in countries like China and India in favor of more manufacturing in Albania (see Table 6.9). For the president of Cotonella S.p.A the Albanian labor market has a high awareness for quality goods and is able to compete with the more experienced Asian countries. As a result, Shqiperia Trikot sh.p.k has gradually acquired a larger share in the total output of Cotonella S.p.A.

The output of Shqiperia Trikot sh.p.k has increased over the years. In 1998, Shqiperia Trikot sh.p.k manufactured only 60% of Cotonella's total output while in 2014 it manufactured 98% of total output, experiencing an increase of 35% compared to 1998. Comparing the level of output in 2015 to that of 2010 there is an increase of 25.92%. In the first years of its operations, Shqiperia Trikot sh.p.k manufactured only intimate apparel for men and women. Later on, it started to manufacture more goods like bras, jersey and items in the product line for children and babies. In the last 5 years, Shqiperia Trikot sh.p.k started to manufacture pajamas for men, women, and children as well as nightgowns. Worth noting is the growth of 366% in the total output of bras from 2010 to 2015. The increase in output of these goods is in full compliance with the head office objective to transfer production from China and India to Shqiperia Trikot sh.p.k. The production cycle in Shqiperia Trikot is presented in Fig. 6.9. In the production cycle assembly occurs in the five subcontractors.

The subcontractors in China manufacture 99% of the cheap bras line of Cotonella S.p.A. It is more convenient to manufacture the economic line bras in China because of the availability of raw materials in the Asian market. The manufacturing of more

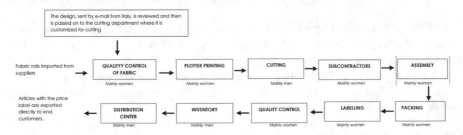

Fig. 6.9 Production cycle in Shqiperia Trikot sh.p.k. *Source* Author's drawing based on the information provided by Shqiperia Trikot sh.p.k

expensive bras lines occurs in Shqiperia Trikot sh.p.k with a total output of 5,000 pieces per year. Pajamas are mostly produced in India because of the specialization this country has in the manufacturing of textiles required for this product range. The general administrator of Shqiperia Trikot sh.p.k states that the increase in the level of output is reflected in the revenues of Shqiperia Trikot sh.p.k. From 2000s and onwards, revenues of Shqiperia Trikot sh.p.k have grown parallel to the expansion in production. In 2015, revenues increased by 126% compared to those in 2000.

One of the most important objectives of Shqiperia Trikot sh.p.k is to achieve a vertical integration system. It aims: (i) to complete transfer of production from Asian countries, (ii) to reduce the level of imported raw materials, (iii) to reduce further the delivery time of finished goods, (iv) to improve the technical skills of employees by bringing them closer to those required by European standards, (v) to realize weaving, stamping, and painting in the city of Shkodra. A vertical production system will be supported by the establishment of a regional center for manufacturing of textiles and fabrics to be used by clothing manufacturing enterprises in South East European countries. Cotonella S.p.A intends to realize an investment of 12 million EUR for this purpose. The regional center is expected to open up a minimum of 800 new jobs. With this investment, Cotonella S.p.A intends to achieve:

- reduction in the time required to manufacture each good;
- reduction in the delivery time of finished products to end customers;
- minimal losses from volatile exchange rates against the euro;
- creation of pool of local suppliers in the provision of raw materials;
- increase in the exports of goods "Made in Albania" so to be able to gain a higher market share.

In order to accomplish this project Cotonella S.p.A is working closely with its subsidiary and Shqiperia Trikot sh.p.k in evaluating the optimal time to make this investment because the regional center needs a minimum output of 5,000 tons cloths to be profitable. Shqiperia Trikot sh.p.k will be in charge to implement the project that would enable to move along the value chain of the textile clothing industry.

Shqiperia Trikot sh.p.k has started the Balkan Peninsula Project. For this project, Shqiperia Trikot sh.p.k has undertaken extensive research to promote the goods of the Cotonella brand in the Balkan Peninsula. Promotion is based on points of sale that are managed by local employees in the countries of the peninsula.

In this project, the head office and Shqiperia Trikot sh.p.k encountered several difficulties including, sale of fake goods, smuggled goods, and competition of cheap goods from Chinese and Turkish brands. In Kosovo, the project has been most successful as customers appreciate comfort and quality. Shqiperia Trikot sh.p.k is also in charge for the administration of the eight shops of Cotonella Brand that are opened throughout Albania. Another project on which the head office is supporting Shqiperia Trikot sh.p.k is on the manufacturing of bathing suits. Shqiperia Trikot sh.p.k has started to manufacture bathing suits thanks to the acquisition of the Linea Sprint enterprise by Cotonella S.p.A.

Assembly tasks in Shqiperia Trikot sh.p.k are fully delegated to its subcontractors (see Table 6.10). Of these subcontractors Madish sh.p.k was established in

Table 6.10 Employment and output in subcontractors of Shqiperia Trikot sh.p.k (2014)

No.	Subcontractor	Establishment date	No. of employees	Process	Goods manufactured	% of output in the total of ShqiperiaTrikot sh.p.k
1.	Madish sh.p.k	1995	103	Sewing Packing	Slip and shirts for men, women, and children	33
2.	Laurus sh.p.k	2006	126	Sewing Packing	Slip and shirts for men, women, and children	22.30
3.	Melkans sh.p.k	2008	88	Sewing Packing	Slip and shirts for men, women, and children	19
4.	Silvana sh.p.k	2012	97	Sewing Packing	Slip for men and women	19
5.	Andrea sh.p.k.	2012	15	Sewing Packing	Slip for men and women	4
6.	Total		429			98

Source Based on the information provided by Shqiperia Trikot sh.p.k

1995 as a division of Shqiperia Trikot sh.p.k. After restructuring occurred in 2001, Madish sh.p.k became a separate entity and a subcontractor of Shqiperia Trikot sh.p.k. Other subcontractors are Laurus sh.p.k (2006), Melkans sh.p.k (2008), Silvana sh.p.k (2012), and Andrea sh.p.k (2012). These enterprises are managed by former employees of Shqiperia Trikot sh.p.k.

Services offered by subcontractors are limited to sewing and packing on all type of intimate apparel (slips, jerseys, bras, pajamas). Among the subcontractors, 50% of the output of Madish sh.p.k and 80% of the output of Laurus is Shqiperia Trikot sh.p.k and the rest is for other customers they work with. The output of the three remaining subcontractors is 100% for Shqiperia Trikot sh.p.k. The five subcontractors have in total 429 employees and all of them are local staff.

Shqiperia Trikot sh.p.k provides support to its subcontractors that consists in:

- selection of employees;
- training of staff working in sewing and packing so they can adapt to the requirements of different customers of Cotonella S.p.A;
- on-site support to employees responsible for quality control;
- technical assistance for the maintenance of machines;
- financial support on investments for subcontractors.

In addition, Shqiperia Trikot sh.p.k continuously monitors each subcontractor. Monitoring occurs at three different levels. Firstly, a technician of Shqiperia Trikot sh.p.k is present throughout the production activities in each subcontractor. Technicians prepare periodic reports identifying problems and difficulties encountered at each production stage. Secondly, the quality control department monitors the standards of the finished goods delivered to Shqiperia Trikot sh.p.k by each subcontractor. Thirdly, Shqiperia Trikot sh.p.k undertakes frequent audits to monitor the implementation of international standards required by Cotonella S.p.A and of technical specifications agreed with each customer. Because of the support offered by Shqiperia Trikot sh.p.k and the increasing market demand for products of Cotonella S.p.A subcontractors have experienced an increase in output and in the number of employees.

The number of employees went down to 78 after sewing and packing services is provided by subcontractors (see Table 6.11). In 2015, the number of female employees Shqiperia Trikot sh.p.k is almost equal to that of male employees. Male employees are present in the department of raw materials while in the cutting department the number of male and female employees is the same. Shqiperia Trikot sh.p.k has 12 employees serving as managers and 66 as non-management staff. Employees having a higher education degree are 24 while those having secondary education are 54. Graduates are placed in leadership positions in Shqiperia Trikot sh.p.k as administrators, heads of departments, or sector managers. The number of female employees in Shqiperia Trikot sh.p.k has fluctuated over the years. Between 1998 and 2000, the company had 385 female employees while in 2004 this number reached to 426. Today, the total number of females working in Shqiperia Trikot sh.p.k is only 34. The majority of female employees previously working for Shqiperia Trikot sh.p.k are now employed by the the five subcontractors. On the other hand, the number of male employees in Shqiperia Trikot sh.p.k was only 10 in 1998, it reached to 66 in 2010 and today it is 44.

Worth emphasizing, is the fact that since the first day of its establishment Shqiperia Trikot sh.p.k has had only Albanian employees. Foreign employees have not been present neither in production nor in management. Foreign staff (mainly Italian) visit Shqiperia Trikot sh.p.k only for short periods to provide technical assistance or to attend executive meetings that aim to monitor the performance of the subsidiary, to specify future objectives, and to identify appropriate expansion strategies.

Employees are selected by the heads of departments and by sector managers. Employees are hired based on previous experience in the clothing industry and the

Table 6.11 Employment according to functions in Shqiperia Trikot sh.p.k (2015)

No.	Sector	Male employees	Female employees	Total
1.	Administration	8	9	17
2.	Raw materials	12	3	15
3.	Cutting	6	6	12
4.	Logistics	18	16	34
5.	Total	44	34	78

Source Based on the information provided by Shqiperia Trikot sh.p.k

skills acquired during the three-month training period. All employees are required to have a secondary education, to speak at least one foreign language, and to have adequate computer skills. The general administrator selects heads of departments and sector managers. They should possess a higher education degree, have relevant experience in the clothing industry, speak at least two foreign languages, and demonstrate good management skills. The policy of Shqiperia Trikot sh.p.k is to select its management staff from employees who have worked in the subsidiary for many years and who have been able to grow in responsibility.

Evaluation of staff is included in daily operations of Shqiperia Trikot sh.p.k. Direct supervisors are responsible to evaluate employees they supervise every three months. Evaluation criteria vary by position and the tasks of employees. For employees placed in low to medium level positions evaluation criteria include commitment, disposability, team-work, and precision in completing the assigned processes. Evaluation of employees having management positions include: reliability, performance, and ability to meet targets. The evaluation form is approved by the head office and is bilingual in English and Italian. Rating for each employee is from 1 to 10. Moreover, Shqiperia Trikot sh.p.k is organized in such a way to ensure daily cross checking of its operations. For the management of the enterprise daily cross-checks are a good indicator for the overall performance of staff and the subsidiary. In the same time Shqiperia Trikot sh.p.k invests financial resources amounting to 30,000 EUR annually for acquiring international audit services on its operations and for the working conditions of its employees.

Moreover, staff training is another engagement of Shqiperia Trikot sh.p.k. Trainings begin on the first day which new employees start a three months training period. During this period employees are on probation, in case they underperform than Shqiperia Trikot sh.p.k reserves the right to replace them. During the three months, experienced employees supervise new employees. Training programs last for 2 weeks, 1 month, 3 months or up to a year. Trainings at the parent company are supplemented with trainings in the premises of Shqiperia Trikot sh.p.k. These trainings are run by domestic and foreign experts and are supported by national universities. Trainings are conducted primarily to:

• improve the quality control of raw materials and finished goods;
• develop new organizational structures of various departments;
• use of advance software to speed up delivery of finished products;

Shqiperia Trikot sh.p.k by the end of 2015 had 20 employees that are trained long term at the head office. In addition to training, employees of Shqiperia Trikot sh.p.k attend ongoing study visits at the head office during which they gain on the spot experience and knowledge. Former employees of Shqiperia Trikot sh.p.k are hired by international brands like Armani and Calvin Klein. Executives of Cotonella S.p.A and those of Shqiperia Trikot sh.p.k emphasize the necessity of a vocational training center to train employees on cutting, design, and quality control with the intend to enrich the local supply of labor in the clothing industry in Albania.

Working hours at Shqiperia Trikot sh.p.k are between 8:00 am and 16.30 p.m. including a 30-min lunch break. The lunch break differs from one department to the

Table 6.12 Compensation of employees in Shqiperia Trikot sh.p.k (2015)

No.	Category of employee	Monthly wage range (EUR)
1.	Sectors managers	420
2.	Personnel performing work with relatively high liability	332
3.	Blue collar employees	260

Source Based on the information provided by Shqiperia Trikot sh.p.k

other. Employees are required to use personal identification card to enter and to leave the company in order to record their working hours. Weekly working hours are in full compliance with the national Labor Code, with a maximum of 50 h per week. Overtime hours are paid 25% more for working on a Saturday, 50% for working on a Sunday. Employees working on national holidays are paid 100% or they can take vacation the next day after receiving the approval of the supervisor. The same policy applies for the five subcontractors.

The annual leave for each employee is 28 working days including Saturdays and Sundays. All employees have 14 days of annual leave on August when Shqiperia Trikot sh.p.k closes its activity. Employees can take the rest of the annual leave depending on their specific needs. Shqiperia Trikot sh.p.k offers daily transportation for all its employees. Staff turnover is very low and even after employees leave Shqiperia Trikot sh.p.k in order to have a new professional experience or go through motherhood they are re-employed in the enterprise. During high intensity production periods, Shqiperia Trikot sh.p.k employs also part time staff.

Compensation for employees is above the average offered by similar enterprises in the clothing industry in Albania. The policy of the enterprise is to periodically review the salaries of employees in accordance with their performance and the number of years working for Shqiperia Trikot sh.p.k. Employees who have more than 10 years working for Shqiperia Trikot sh.p.k receive an additional salary (see Table 6.12).

6.3 Naber Konfeksion sh.p.k

This section presents the second case study of the research. It concentrates on the operational and production activity of the German enterprise Naber Moden and its subsidiary Naber Konfeksion sh.p.k[5] that is composed of three production plants.

Presentation of the case starts with a description of the host region, the activity of the group and it continues with the activities occurring in the subsidiary.

---

[5]Naber Konfeksion sh.p.k is composed of three production plants Naber Konfeksion 1 sh.p.k, Naber Konfeksion 2 sh.p.k, and Grori Konfeksion sh.p.k.

6.3.1 The Region of Durres

The region of Durres is situated in the west of Albania, 35 km from Tirana, the capital
of Albania (see Fig. 6.10). The region of Durres is the second largest in Albania in

Fig. 6.10 The map of Albania with the region of Durres circled. *Source* www.europe-atlas.com

Table 6.13 Table presenting the region of Durres (2017)

Indicator (2017)	Region of Durres	Albania
Population	284,823	2,876,591
Total GDP in (Mln/EUR)	1.147	11,564
GDP per capita in (EUR)	3,994	4,024
GDP composition in (%) of the total GDP		
Agriculture, forestry, and fishery (%)	17.80	22.70
Industry (%)	17.90	13.90
Construction (%)	11.60	10.20
Trade, transport, hotels, and restaurants (%)	24.80	18.10
Information and communication (%)	1.90	3.50
Financial and insurance activities (%)	1.60	2.80
Real estate activities (%)	9.40	6.60
Scientific, professional, administrative, and supporting activities (%)	5.30	6.70
Public administration, education, health, and social activities (%)	8.10	12.50
Artistic, entertaining, and related activities (%)	1.80	3.00
Number of enterprises	12,963	162,452
Number of foreign owned enterprises	430	3,704

Source INSTAT, and the Chamber of Façon in Albania

terms of population and economic activity. In 2017, the region had a population of 284,823 inhabitants. It has the largest seaport in Albania through which over 1,500,000 passengers commute annually and where over 3 million tons of goods circulate within a year.

In 2017, the GDP in current prices for the region of Durres amounted to 1147 million EUR having a share of 9.91% in the country's GDP (see Table 6.13). In the same time, the GDP per capita for the region was 3,994 EUR at the market exchange rate. Trade (wholesale and retail) and tourism related economic activities (restaurants, hotels, and entertainment) have the highest share with 24.80% of the region GDP.

Agriculture is one of the main economic activities in the region spread mainly in rural areas. Most agricultural products are exported to the European market. This region has also a well-developed fish processing industry, which is based on the variety of fish and related species that are found in the Adriatic Sea.

In the early 1990s, with the establishment of the market economy, a high number of clothing manufacturing enterprises settled in the region of Durres to benefit from the presence of the largest port and the proximity with the European market (especially to Italy). Moreover, in 2018, in the region of Durres were 12,921 active enterprises making 7.93% of the total active enterprises in the country (see Fig. 6.11). Of the

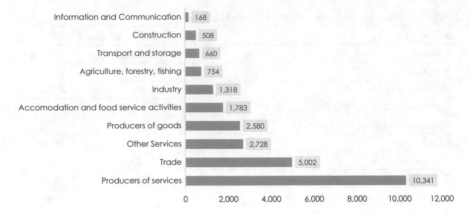

Fig. 6.11 Number of active enterprises in the region of Durres (2018). *Source* National Statistical Institute (INSTAT), Business Register 2018

active enterprises 3,690 have female owners. In the region 487 enterprises have foreign ownership while 246 have are jointly shared between foreign and local citizens. The region of Durres is dominated by small and medium size enterprises. Among active enterprises, 11,013 of them employ up to four persons while those employing up to 49 employees are 748. Enterprises with over 50 employees are only 236.

6.3.2 The Naber Moden Group

This section introduces the Naber Moden Group with reference to its history and operating activity. In presenting the activity of the group this section starts with the history of the group, the types of articles manufactured, the customers, suppliers, and the organizational structure.

As of 2015, Naber Moden is composed of: (i) the head office in Germany, (ii) 8 stores located in the Bavaria region, (iii) three production plants in Albania, (iv) subcontractors in FYR Macedonia.

Naber Moden was established in 1951 when Herbert Naber opened in Nuremberg, Germany, a clothing manufacturing enterprise named "Naber Damen Moden", specialized in female outfits and having only six employees (see Fig. 6.12).

In the 1990s, the enterprise shortened its name to Naber Moden as the range of its products was not limited to those designated for women. Until the end of 1960s, Naber Moden served as a subcontractor for production of blouses to its German customers. In 1970s, Naber Moden transferred production of blouses to Greece where it operated under rented facilities with no purchases in land or buildings. After almost ten years of production, when Greece became a member of the European Union in 1981, Naber Moden experienced rising manufacturing costs coming mainly from the increase in the cost of labor forcing the enterprise to move production towards Eastern Europe

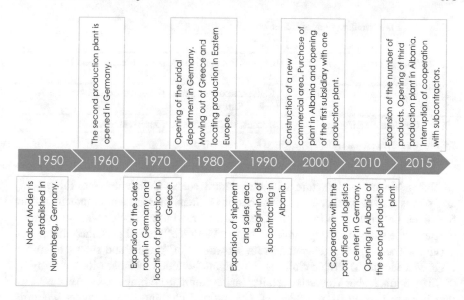

Fig. 6.12 Main events in the history of Naber Moden. *Source* Based on the information provided by Naber Moden

where the labor force was cheaper. It started to manufacture through subcontractors in Poland, Bulgaria, and FYR Macedonia. It operated in Poland for almost 6 years while it maintained short-term contracts for up to 2 years with subcontractors in Bulgaria and FYR Macedonia. In 1995, Naber Moden started to produce in Albania by subcontracting Ambra sh.p.k in the city of Durres. In 2015, after opening three production plants, Naber Moden terminated outsourcing of assembly activities to five local subcontractors in Albania.

Table 6.14 summarizes the main characteristics of production plants possessed by Naber Moden in Albania.

Table 6.14 Production plants owned by Naber Moden in Albania

Naber Moden production plants in Albania			
Plant	Naber Konfeksion 1	Naber Konfeksion 2	Grori Konfeksion sh.p.k
Products	Blouses, parkas, gilets, skirts, dresses, jackets, shirts, etc.	Trousers, parkas, gilets, skirts, dresses, jackets, shirts, etc.	Jeans
Jobs (June, 2015)	400	320	80
Industrial facilities m² (June, 2015)	8,500	1,141	850
Date of setting up	2004	2012	2015

Source Based on the information provided by Naber Konfeksion 1 sh.p.k

Table 6.15 Main clothing lines of Naber Moden

No.	Main clothing lines	Articles
1.	Young fashion	Blouses, trousers, shirts, skirts, dresses, jackets, coats, parkas, furs, etc.
2.	Women fashion until size 54	Blouses, trousers, shirts, skirts, dresses, jackets, coats, parkas, furs, etc.
3.	Children clothing line	Blouses, shirts, dresses, trousers, skirts, coats

Source Based on the information provided by Naber Moden

Naber Moden manufacturers its own brand Naber Collection (see Table 6.15) and serves also as a subcontractor for major German international brands including S. Oliver, Lebek International, Brax, etc. Own brand accounts for 60% in the total output while subcontracting for 40%. Naber Moden offers full package with design services. The output is generated in its subsidiary in Albania and subcontractor in FYR Macedonia. The main clothing lines manufactured by Naber Moden[6] are:

In addition, Naber Moden is a retailer for ceremonial clothing lines and accessories that it buys directly from producers. Ceremonial clothing lines include: wedding dresses and male gowns, communion clothing, shoes and bags, accessories (belts, scarfs, hats, bracelets, earrings).

Naber Moden sells its output mostly in Germany. Articles manufactured for Naber Collection are sold through the eight shops located in the Bavaria region while the rest is delivered to respective customers. The number of customers of Naber Moden has increased over the years.[7] The customers of Naber Moden are mainly international German retailers (see Table 6.16). Naber Moden produces finished articles for Brax International[8] (trousers), for Otto Group,[9] for Weber/Toni[10] dress. Most of the children line manufactured by Naber Moden is for the well-know Mafrat group, which includes brands like Gianfranco Ferre, Ferarri, Laura Biaggiotti, etc. During an interview in June 2015 with Bern Naber he mentioned that for a pair of trousers Brax pays 3–4 EUR for the cut, make and trim service. Brax sells in its shops a pair of trousers for 100–150 EUR.

---

[6]During an interview realized with Bernd Naber in April 2015 he stated that Naber Moden is working towards obtaining ISO standards in clothing manufacturing In order meet the criteria to receive the international standards Naber Moden has received consulting and auditing services from international companies and extensive feedback from its customers.

[7]In April 2015, during an interview with the founder Herbert Naber noted that "Even during the international economic crisis of 2008–2010 the enterprise continued not only to take orders from its customers but also expanded the range of its customers".

[8]Brax international has more than 1,720 shops in Germany and 115 stores worldwide (www.corporate.brax.com).

[9]Otto Group consists of 123 companies in more than 20 countries in Europe, Americas and Asia. In e-commerce the group is the number one fashion and life style company (www.ottogroup.com).

[10]Weber/Toni has more than 1,000 company managed stores, 2800 shops-in-shops, and 281 franchise stores (www.gerryweber.com).

Table 6.16 Main customers of Naber Moden and daily output (pieces) (2015)

No.	Main customers	Type of articles manufactured	Average daily production (pieces)
1.	Brax	Trousers	500
2.	Adler	Blouses, dresses	370
3.	Klingel	Blouses, dresses, skirts, shirts, trousers, parka, gilet	250
4.	Bader	Blouses, dresses	250
5.	Emilia lay	Blouses, shirts	250
6.	K & L Ruppert	Blouses, tops, shirts	250
7.	Unit O	Blouses, dresses, dirndl	300
8.	Otto Baur Witt WeidenSieh Ana	Blouses, skirts, trousers, jackets, parka, shirts	300
9.	Katag	Blouses, dresses	200
10.	Weber/ Toni Dress	Blouses, tops, shirts	250
11.	Gelco	Blouses, shirts, jackets	600
12.	BG Mode	Dresses, dirndl[a], blouses, shirts	200
13.	Mafrat	Blouses, shirts, dresses, skirts, coats, parka, for children	150
14.	Versace	Dresses, trousers, skirts	200
15.	Vestebene	Trousers	150

[a]Traditional German dresses that are popular during October Fest
Source Based on the information provided by Naber Moden

Naber Moden acquires all of its raw materials from foreign suppliers. Local suppliers are absent for two main reasons: (i) there are no local suppliers able to produce the variety of fabrics used by Naber Moden in production of cloths and (ii) local suppliers are unable to meet the standards required from customers even for simple raw materials like cardboard boxes and labels.[11] According to Bernd Naber "Everything is imported". In 2015, the quantity of imported fabric was 1,574,000 linear meters coming mainly from Turkey and Egypt. From Turkey are imported mostly fabrics containing a high degree of cotton while from Egypt it imports more delicate fabrics like silk, etc.

For many years, Naber Moden has been working with the same pool of suppliers. According to Bern Naber, a long and good cooperation is essential to guarantee the quality and the on-time delivery of raw materials. Occasionally, Naber Moden asks its suppliers to produce special fabrics when customers are launching in the market a new product or a limited collection. Special fabrics are mostly produced in Egypt as

---

[11]According to an interview wit Bern Naber on raw materials in June 2016 he states the "Everyone is important".

Naber Moden is working with a large regional manufacturing enterprise specialized in stamping which meets the technical specifications required for production in Naber Moden. Selection of suppliers and placements of orders occurs in the head office while import procedures are performed in the subsidiary.

The head office is also responsible for organizing annual and semiannual meetings with suppliers. These meetings serve to: (i) revise the schedule of orders, (ii) recommend technical improvements in existing fabrics, (iii) introduce new fabrics that need to be included in upcoming orders, (iv) and to be informed on the projects suppliers are working on that may be of interest to Naber Moden.

Naber Moden applies a rigid quality control system on raw materials. Initial quality control consists on numerous tests on sample fabrics including endurance, stretching, appearance retention, thermal resistance. These tests are performed in the head office and are mandatory before making any order from suppliers.

A second level of quality control is performed before and during production in the plants located in Albania. This second level of quality control consists in the identification of possible marks, holes, missing threads, etc.

Naber Moden has subcontracted an Albanian company specialized in international transport services to deliver finished articles manufactured in Albania to storage facilities that are managed by the head office in Germany. The storage facilities operate with an inventory system, which records the inflow/outflows of goods and shows the stock at any point of time. Naber Moden is not responsible to deliver goods to every customer.

When orders are placed in storage facilities customers like Brax, Gerry Weber are informed to take them out. Under specific circumstances, customers can retrieve finished goods from Naber Collection shops that are spread in the Bavaria region in Germany or to end customers.

Today, the head office of Naber Moden is still in Nuremberg, Germany and the enterprise continues to be a family run business. It has 85 employees in Germany. The head office of Naber Moden operates under the organizational structure presented in Fig. 6.13 is composed of high management and six departments required to run the enterprise.

Fig. 6.13 Organizational structure in the head office of Naber Moden in Germany. *Source* Author representation based on the information provided by Naber Moden

6.3.2.1   High Management

It consists of the founder, Herbert Naber and his two sons. The founder serves as the honorary president and frequently monitors production in Albania. He is assisted mostly by his son Bernd Naber a textile engineer who also leads the top management.

6.3.2.2   Research and Development

It is a department that has five employees, with one of them serving as the department head. The main responsibilities of the department are to: (i) monitor the global trends of new raw materials used in the clothing industry, (ii) identify new categories of fabrics that may be used in production, (iii) propose new specifications to suppliers according to the requirements of customers, and (iv) expand the Naber Collection by including new articles.

6.3.2.3   Design

It is a department in the head office run by eleven employees. This department is responsible for: (i) preparing for production the designs received by customers on samples and articles, (ii) proposing new designs to customers on their articles, and (iii) creating the designs for the Naber Collection.

6.3.2.4   Customer Relations

It is a department that has five employees and it is in charge to obtain continuous feedback from customers in order to identify new services that can be introduced to customers.

6.3.2.5   Raw Materials

It is a department run by one employee who is in charge for getting hold of raw materials in accordance with technical specifications obtained from customers and the design department. The employee is also responsible for ensuring that raw materials are delivered to production plants in Albania according to the order placed by Naber Moden for its own brand and respective customers.

6.3.2.6   Logistics Department

It is a department that is run by 10 employees who are responsible for the delivery of finished articles to end customers and for the placement of Naber Collection in the eight stores owned by the enterprise in the Bavaria region of Germany.

6.3.2.7   Sales

It is a department with the highest number of employees and it is in charge for all sales in Naber Moden. The department is composed of two main units one that is in charge for the sales of Naber Collection and the other one that is in charge for the sales of the customers of Naber Moden.

From the organizational structure of Naber Moden the head office still remains responsible for key segments of the clothing manufacturing value chain including research and development, design, and relationships with customers. In Albania, the three production plants of Naber Moden undertake only assembly functions. In June 2015, during an interview with Bernd Naber he said that in Germany is done: "Everything except sewing".

Figure 6.14 presents the value added activities in the clothing industry according to value added activities presented in Chap. 3. In blue are value adding activities occurring in the head office in Germany (except production) and in white production occurring in Albania.

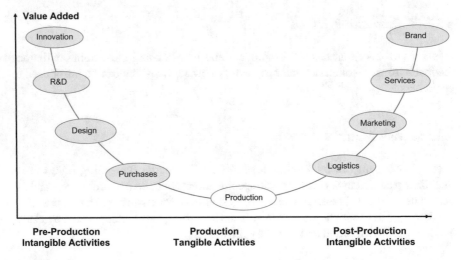

Fig. 6.14 Value adding activities in Naber Moden. *Source* Based on the information provided by Naber Moden

6.3.3 The Subsidiary Naber Konfeksion sh.p.k Clothing Manufacturing Enterprise

This section presents the production activity of Naber Moden plants in Albania. The information included in this section was obtained from fieldwork in the three production plants in Albania.

In the 1990s, the management of Naber Moden feared that the labor cost in Eastern European countries would increase as previously occurred in Greece in the early 1980s. Therefore, the owners of Naber Moden started to look for new potential territories where to locate production. In 1995, Naber Moden entered in the Albanian territory and started production through its first subcontractor Ambra sh.p.k.

As Naber Moden extended its presence in Albania the number of subcontractors expanded to five. For Bernd Naber Albania was considered a good option as it still has the lowest labor costs in the region while production costs in Asia were going up due to the increase in the price of raw materials, transportation, and labor costs. Gradually, Naber Moden started to receive more orders from customers in German speaking countries in Europe (Germany, Austria and Switzerland) that opened up new opportunities for the enterprise to have its own production site in Albania and not to depend entirely on local enterprises that served as subcontractors.

In 2004, Naber Moden opened its first plant in Albania by acquiring production facilities of Trumpf Blusen,[12] another German company located in the city of Durres, which was experiencing numerous losses. The acquisition led to the establishment of the first plant Naber Konfeksion 1 sh.p.k (see Fig. 6.15).

Before opening a second plant in Albania, the high management in the head office considered other countries like Romania, Macedonia and Turkey, which resulted inappropriate for locating production. In Romania, Naber Moden could lose workers due to emigration, in Macedonia they could not hire more people as they are already employed, and in Turkey, the labor cost was 1,000 euro/month compared to 300 euro/month in Albania. Consequently, the owners decided to open in 2012 a second production plant, Naber Konfeksion 2 sh.p.k, specialized mostly in the production of trousers.

In 2015, a third production plant on rented facilities for, Grori Konfeksion sh.p.k, became operational. It is specialized in production of jeans for Primark and Brax. These international brands decided to transfer production of jeans from Turkey to Albania in order to achieve shorter delivery times and to save on labor costs.

To continue, the level of employment in the three plants of Naber Moden in Albania has increased (see Table 6.17). In 2004, Naber Konfeksion sh.p.k employed

---

[12]Founded in 1932 in Munich traditional company Trumpf Blouses Dresses GmbH & Co. KG was temporarily the largest blouses manufacturer in Germany. The company struggled for years with quality problems and declining customer acceptance in early 2000s, which had continuous sales declines result. In 2005, the year of bankruptcy, with annual sales of only 28 million EUR and the workforce had now been reduced to 150 employees Today, the status of the company is described as completed insolvency proceedings (www.mhbk.de).

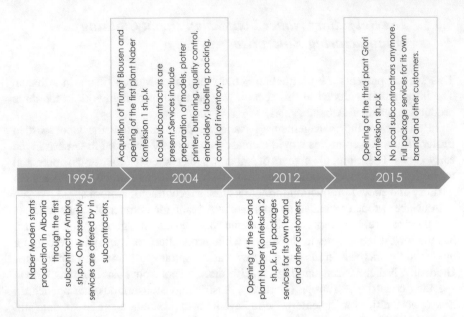

Fig. 6.15 Main events in the history of Naber Konfeksion sh.p.k. *Source* Based on the information provided by Naber Moden

Table 6.17 Level of employment in Naber Konfeksion sh.p.k in Albania (2004–2015)	No.	Year	Level of employment
	1.	2004	190
	2.	2009	320
	3.	2012	580
	4.	2015	80

Source Based on the information provided by Naber Konfeksion sh.p.k

in Albania 190 workers while in 2015 the number of workers reached to 800, an increase of 321% and of 710 workers compared to 2004.

The plants of Naber Konfeksion sh.p.k in Albania are organized in such a way to offer full package services (see Fig. 6.16). The organizational structure in the three plants is shown below. The general manager, the administrator, the production manager, and the German technicians are in charge to oversee production in the three plants. The manager in charge for production supervises and visits daily the responsible for the production division in the three plants. Moreover, the subsidiaries have in common the finance, the logistics, and the inventory department (see Table 6.18). The production departments in the three plants cooperate closely with each other as some processes for specific products like trousers cutting occurs in the plant while assembly in another plant. In addition to production, the two smaller plants work closely

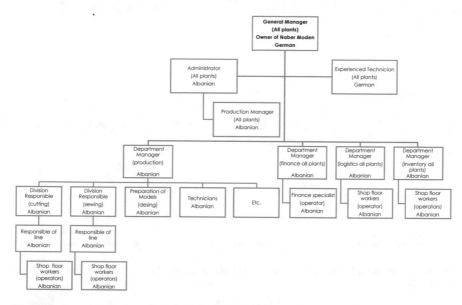

Fig. 6.16 Organizational structure in Naber Konfeksion sh.p.k plants in Albania. *Source* Author representation based on the information provided by Naber Konfeksion sh.p.k

Table 6.18 Production units of Naber Konfeksion sh.p.k plants in Albania

No.	Subsidiary	Divisions of production department
1.	Naber Konfeksion 1 sh.p.k	Preparation of models, plotter, cutting, sewing, buttoning, labeling, ironing, packing
2.	Naber Konfeksion 2 sh.p.k	Plotter, cutting, sewing, buttoning, labeling, ironing, packing
3.	Grori Konfeksion sh.p.k	Sewing, buttoning, labeling, ironing

Source Based on the information provided by Naber Konfeksion sh.p.k

with Naber Konfeksion sh.p.k for the distribution of raw materials and transportation of finished articles.

The organizational structure of departments is broken down into divisions, production lines, and shop floor workers that serve as operators. The three production plants in Albania differ in the divisions available in the production department:

Figure 6.17 presents the value added activities in the clothing industry according to value added activities in global value chains presented in Chap. 3. In dark are the value adding activities occurring at the production plants in Albania. These activities are limited only to production.

This section presents the level of investments Naber Moden has made in its subsidiary Naber Konfeksion sh.p.k in Albania (see Fig. 6.18). Investments are categorized into those made on operating facilities, on machines required to manufacture finished goods, and on information technology.

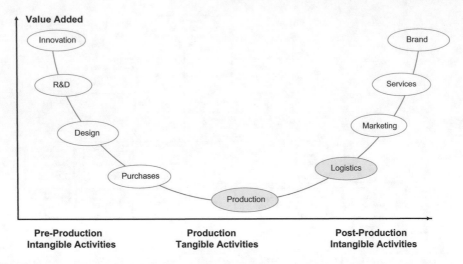

Fig. 6.17 Value added activities of Naber Konfeksion plants in Albania. *Source* Naber Konfeksion sh.p.k

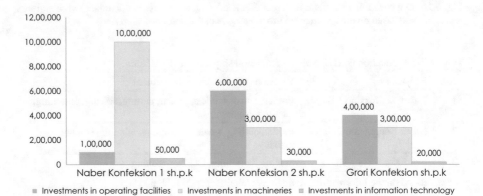

Fig. 6.18 Investments in Naber Konfeksion sh.p.k (2015). *Source* Based on the information provided by Naber Konfeksion sh.p.k

Operating facilities in Naber Konfeksion 1 sh.p.k occupy a surface area of 8500 m^2 including production site and residential area (see Table 6.19). These facilities are fully owned by Naber Moden. Initially, Naber Moden invested 100,000 EUR in renovating production site, which was previously occupied by Trump Blusen. Additional investments transformed Naber Konfeksion 1 sh.p.k into a two store production facility. An amount of 50,000 EUR is spent in renovating the two guest houses located near the production site. These guesthouses host the high-level management, the technicians and the international consultants.

In 2012, Naber Moden inaugurated its second production plant in Albania, Naber Konfeksion 2 sh.p.k. The facilities of Naber Konfeksion 2 sh.p.k occupy an area of

Table 6.19 Investments in Naber Konfeksion sh.p.k (2015)

Category of investment	Amount in EUR
Investments in operating facilities	
Naber Konfeksion 1 sh.p.k	100,000
Naber Konfeksion 2 sh.p.k	600,000
Grori Konfeksion sh.p.k	400,000
Investments in machines	
Cutting machines	100,000
Sewing machines	1,100,000
Embroidery machines	100,000
Specialized machines (buttoning, studding, etc.)	150,000
Ironing machines	50,000
Pressing machines	100,000
Investments in information technology	
Hardware	70,000
Software	30,000
Total of investments	2,800,000

Source Based on the information provided by Naber Konfeksion sh.p.k

1141 m^2 and are fully owned by Naber Moden. Even though smaller in size than the first plant, the head office has made an investment of over 600,000 EUR in the facilities of Naber Konfeksion 2 sh.p.k.

In 2015, Naber Moden opened a third production plant in Albania. The facilities are rented and occupy an area of 850 m^2. Even though under rented facilities, Naber Moden has invested 400,000 EUR in adapting facilities for production. A common feature in the operating facilities of the three plants is the availability of parking space for: (i) placement of large trucks used to transport finished articles, (ii) positioning of shuttle buses used for transportation of employees, and (iii) parking places for the administration.

To continue, Naber Konfeksion sh.p.k has acquired a variety of machines needed to realize production in Albania. In total, there are six cutting lines[13] available in three production plants in Albania. Samples are prepared for cutting by two new plotters purchased by Naber Konfeksion sh.p.k as specialized machines for 30,000 EUR in 2012.

Sewing machines are categorized depending on the type of articles they are used to produce and include machines for (i) coats, parkas, and jackets, (ii) shirts and skirts, blouses, (iii) trousers, and (iv) for manufacturing traditional German dresses

---

[13]Standardized garments like shirts and skirts prepared are cut in large quantities in the automated machines while delicate fabrics used mostly in the production of blouses go through the manual cutting machines.

"dindl". Most of sewing machines can be used to manufacture different articles as they operate on a program that allows employees to switch articles. In addition, machines used to perform specific processes in manufacturing of cloths include:

- automated embroidery[14] machines used to prepare the design in different articles; machines used to prepare ruffles on a variety of fabrics and for different types of articles;
- two lines of buttoning, stitching, and studding machines (Buttoning and stitching machines are used on any kind and size of buttons and models of stitches);
- automated studding machines able to insert studs in any fabric and kind of cloth;
- ironing machines that operate based on vapor irons and ironing tables which can be used uninterruptedly as they are connected to large heaters that ensure hot water at all times.

Up to 2015, the amount invested in machines required for production amounts to 1,300,000 EUR or 62.5% of the total investment. When machines require spare parts they are purchased in Germany.

Apart from investing in machines, Naber Konfeksion sh.p.k has also invested in information technology (IT). It has invested an amount of 80,000 EUR in hardware which include desktops, laptops, and printers while engineers employed by Naber Moden use working station computers. The main IT investments are: (i) an advanced software purchased to run the plotters, (ii) the "Gemini software" used to prepare various samples based on the design received from the head office, and (iii) an inventory program which assists the inventory and the finance department to keep track between new orders placed by customers and those that are ready for shipment.

This section presents the production activity in Naber Konfeksion sh.p.k including the production cycle, articles manufactured, and former subcontractors of Naber Konfeksion sh.p.k.

Production in Naber Konfeksion sh.p.k is divided in two stages. The first stage occurs in the head office and the second takes place in production plants located in Albania (see Fig. 6.19). Production starts within the design department in the head office where employees prepare for manufacturing the design[15] of the requested articles including those for customers and Naber Collection.

After, the design is finalized employees in the raw materials department perform quality tests (stretching, de-coloring, heat resistance, etc.) on the fabric chosen for a particular model. Then the sales department sends electronically the design to the production plants in Albania and with courier, the amount of fabric required to produce two samples of the article. Together with the fabric is included the model

---

[14]Embroidery machines operate based on the model inserted through the USB port in the computer connected to machines. In the computer monitor employees monitoring follow every step until completion of the model inserted through USB port is finalized. In the monitor also appear the various signs that indicate when the embroidery process is interrupted like in the case when a machinery is running out of threads or when a needle is not functioning properly.

[15]Employees can either receive the design directly from customers or can propose to them various designs and may work for up to a week to finalize it. With regard to Naber Collection, design is prepared within the department and is approved by the top management in the head office.

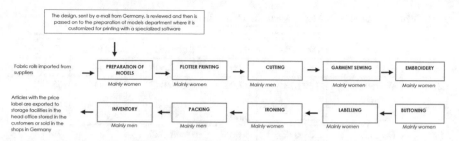

Fig. 6.19 Production cycle (Embroidery is applicable only to specific articles manufactured in the subsidiary. The dotted line indicates a production cycle without embroidery) in the plants of Naber Konfeksion sh.p.k in Albania. *Source* Author representation based on the information provided by Naber Konfeksion 1 sh.p.k

card that contains the technical specifications and the deadline when the head office expects the delivery of samples.

Within production plants in Albania, engineers working in the preparation of model department print the designs received electronically from the head office through plotters (specialized printers). The printed design together with the fabric is cut and then undergoes sewing, embroidery (if applicable), buttoning, and labeling. Finished samples get specialized ironing depending on the type of fabric, are properly packed, and receive a final control with the inventory departments before reaching the head office. In case a customer or the top management for Naber Collection decides to make further changes on the samples than the whole production cycle is repeated. This process may take up to two weeks. If samples are satisfactory then an order is placed specifying the quantity and the delivery time. The same production cycle is repeated until the entire quantity is produced.

To continue, articles manufactured in the plants of Naber Konfeksion sh.p.k are diverse. Within four years, overall production has experienced an increase of 51.77% (see Table 6.20). The increase is mostly attributed to production of blouses and trousers coming mainly from the output generated after the opening of the second plant, Naber Konfeksion 2 sh.p.k. Fur articles, gilets, and the line of children have the lowest share in output of Naber Konfeksion sh.p.k.

Naber Moden has cooperated with Sequa Consulting Company, a partner of the German Ministry of Economic Cooperation, for improvement in production. Thanks to this program, Naber Moden and Naber Konfeksion sh.p.k managed to optimize various production processes and to introduce additional operations in its production plants. In addition, Naber Konfeksion sh.p.k has also worked with 30 consultants from Weis Consulting, a German consulting company specialized in the clothing industry, and with the Polytechnic University of Tirana. From this cooperation, the head office and Naber Konfeksion sh.p.k have been able to improve production within its plants by:

- organizing work production in line with the placement of carriages that move from one department to; another to complete all production cycle;
- registering the time required to complete each process;

Table 6.20 Output of Naber Konfeksion sh.p.k in Albania (2010–2015) (pieces)

No.	Article	Output 2010 (articles)	As % in the total	Output 2015 (articles)	As % in the total	Change in %
1.	Skirts	35,125	5.00	42,646	4.00	21.41
2.	Women trousers	189,673	27.00	329,843	30.94	73.90
3.	Shirts	35,125	5.00	52,984	4.97	50.85
4.	Dresses	21,075	3.00	27,323	2.56	29.65
5.	Blouses (long/short sleeve/ruffles)	280,997	40.00	420,457	39.44	49.63
6.	Jackets	21,075	3.00	28,323	2.66	34.39
7.	Coal	21,075	3.00	30,984	2.91	47.02
8.	Parka	28,100	4.00	37,646	3.53	33.97
9.	Gilets	21,075	3.00	25,984	2.44	23.30
10.	Fur articles	14,050	2.00	17,661	1.66	25.71
11.	German traditional outfits (Dirndl)	21,075	3.00	29,969	2.81	42.20
12.	Children Line	14,050	2.00	22,323	2.09	58.88
13.	Total output	702,493	100	1,066,142	100	51.77

Source Based on the information provided by Naber Konfeksion 1 sh.p.k

- implementing a quality management system for each process starting from inception up to packing;
- counting on new planning techniques to meet relevant deadlines;
- speeding ironing of finished goods.

In 1995, the first subcontractor of Naber Moden in Albania was Ambra sh.p.k. After three years, Naber Moden started to expand its subcontractors by adding Dyrrah Sped sh.p.k (1998) and Nels sh.p.k (1999). As sales were going up, Naber Konfeksion engaged two more subcontractors Ones sh.p.k (2000) and Shiroka sh.p.k (2001) (see Table 6.21). In 2015, the owners of Naber Konfeksion sh.p.k decided to end cooperation with five subcontractors. Subcontractors of Naber Konfeksion sh.p.k were all Albanian enterprises employing only local staff. Even though they provided the same services (cutting, sewing, and packaging), their degree of specialization was different.

Three subcontractors specialized in manufacturing of blouses, one in manufacturing of coats and shirts, and the remaining only in trousers. The quantity manufactured by each subcontractor was between 1,000 and 1,500 pieces a week with a maximum amount of 6,000 pieces per subcontractor a month. According to Bern Naber cooperation with subcontractors ended as "To do it on your own is to have it

Table 6.21 Subcontractors of Naber Konfeksion sh.p.k in Albania

No.	Subcontractors	Date of cooperation	End date of cooperation	Number of employees	Processes	Articles manufactured	Level of production
1.	Ones sh.p.k	2000	2015	40	Cut, Sewing, Ironing and packaging	Blouses	1,000–1,500 pieces in a week
2.	Shiroka sh.p.k	2001	2015	60	Cut, Sewing, Ironing and packaging	Blouses	1,000–1,500 pieces in a week
3.	Ambra sh.p.k	1995	2015	300	Cut, Sewing, Ironing and packaging	Coats, shirts	1,000–1,500 pieces in a week
4.	Dyrrah Sped sh.p.k	1998	2015	70	Cut, Sewing, Ironing and packaging	Trousers	1,000–1,500 pieces in a week
5.	Nels sh.p.k	1999	2015	50	Cut, Sewing, Ironing and packaging	Blouses	1,000–10,500 pieces in a week

Source Based on the information provided by Naber Konfeksion 1 sh.p.k

under control".[16] During cooperation with its subcontractors Naber Konfesion sh.p.k provided on-going support to its subcontractors that consisted in:

- training of employees to meet technical specifications required by various customers;
- periodic training of employees responsible for the quality control during production;
- technical assistance in increasing operational efficiency;
- training of female employees in mastering sewing and cutting tasks.

In addition, Naber Konfeksion sh.p.k monitored production in each subcontractor. Monitoring occurred mostly through experienced technicians coming from the head office that periodically visited each subcontractor. Technicians prepared reports informing the top management of Naber Konfeksion sh.p.k on the difficulties and problems encountered during production. Furthermore, the management of Naber Konfeksion sh.p.k periodically met administrators of each subcontractor to discuss on: (i) the degree of performance for a given period, (ii) possible solutions to overcome operational difficulties, and (iii) the degree of compliance on technical specifications imposed by Naber Konfeksion sh.p.k. During on site visits and interviews

---

[16]Interview in June 2015 in the premises of Naber Konfeksion sh.p.k.

with representatives of Ambra sh.p.k and Ones sh.p.k they stated that cooperation
with Naber Moden helped them to improve their production capabilities and to continue manufacturing even though cooperation with Naber Konfeksion sh.p.k has terminated. Today, these subcontractors offer assembly services directly to customers
of Naber Konfeksion sh.p.k.

Naber Moden employs 800 workers in Albania of which 400 work in Naber Konfeksion 1 sh.p.k (see Table 6.22). Only 5.25% of employees are male compared
to 94.75% of females. Assignment of responsibilities between male and females
employees is done based on the tasks that are to be completed. Tasks that require attention to detail like sewing and ironing are completely occupied by female employees
while in tasks that require physical strength male employees have the lead.

Technicians, drivers, electricians, gardeners, and cleaners assist in the daily operations in the three plants. Among them, some technicians have over 20 years of
experience in maintaining machines and are able to fix on the spot the majority of
problems occurring during production. In addition, technicians are capable to identify
the parts needed to be replaced without causing long interruptions in the production
cycle. Cleaners continuously remove the waste generated after cutting and sewing.

Employees of Naber Moden in Albania are only local staff. German experts and
technicians that periodically support local employees, are employed by the head
office and are in charge for: (i) providing on-going technical assistance, (ii) suggesting
new techniques for increasing labor productivity, (iii) participating in administrative
meetings on the performance of operations and the level of output and (iv) setting
future targets for production in Albania. The high management of the head office
visits Albania every month to evaluate the performance and to provide solutions on
problems that could not to be solved by local administration.

Criteria for the selection of employees depends on the degree of responsibility
and the difficulty of assigned tasks. Local administrators are directly hired from the
owners of Naber Konfeksion sh.p.k. They should have a university degree, speak at
least two foreign languages, and have a good command of computer programs. Heads
of departments, divisions, and responsible for production lines are hired from local
management depending on previous experience in the clothing industry. They need to
have a secondary or higher degree, speak at least one foreign language, and possess a
basic level of computer capabilities. Engineers working for Naber Konfeksion sh.p.k
must have university degrees in textile engineering and previous experience in the
clothing industry.

In Naber Moden, employees have the opportunity to get promoted. For example, employees that start working in sewing lines after several years can become in
charge of a division or unit. According to the management, when it comes to hiring
shop floor employees in charge for cutting, sewing, and buttoning the enterprise has
difficulties in finding employees able to produce quality goods and willing to learn
new processes and techniques. For Naber Konfeksion sh.p.k it is also hard to find
additional female employees. New female employees are found through existing
employees or via advertising in the surroundings of manufacturing plants. Support
from regional employment centers has been very limited in finding female employees.
Naber Moden has also experienced difficulties in identifying middle management

Table 6.22 Employment in Naber Moden plants in Albania (2015)

No.	Category	Naber Konfeksion 1 sh.p.k			Naber Konfeksion 2 sh.p.k			Grori Konfeksion sh.p.k			Totals
		Total	Female	Male	Total	Female	Male	Total	Female	Male	Value
1.	Administration	5	4	1	4	4	0	2	2	0	11
2.	Finance	2	2	0	0	0	0	0	0	0	2
3.	Plotter	3	3	0	2	2	0	0	0	0	5
4.	Inventory	2	0	2	2	0	2	0	0	0	4
5.	Labelling	3	1	2	2	0	2	2	0	2	7
6.	Cutting	22	16	6	15	5	10	0	0	0	37
7.	Buttoning	10	10	0	12	12	0	0	0	0	22
8.	Preparation of models	5	4	1	0	0	0	0	0	0	5
9.	Ironing	15	15	0	15	15	0	9	9	0	39
10.	Packaging	12	12	0	10	10	0	0	0	0	22
11.	Sewing lines	210	210	0	200	200	0	52	52	0	462
12.	Maintenance	12	5	7	7	3	4	4	2	2	23
13.	Logistics	1	0	1	1	0	1	1	0	1	3
14.	Employees in the training period	98	98	0	50	50	0	10	10	0	158
15.	Totals	400	381	19	320	302	18	80	75	5	800

Source Based on the information provided by Naber Konfeksion1 sh.p.k

employees that possess adequate knowledge of the industry and are interested in joining the enterprise.

Evaluation criteria in the subsidiary vary by position, responsibility, and required experience. For low to medium level ranked employees evaluation criteria are commitment, teamwork, precision, and the time required in manufacturing the assigned quality of articles. Middle management is evaluated on the ability to instruct and guide their employees, to meet targets, and to improve staff performance. The evaluation form is prepared in the head office and is approved by the high management of Naber Konfeksion sh.p.k. The form is structured in such a way to incorporate production cycle and quality cross-checking on a daily basis.

Naber Moden has training programs for all its employees working in Albania. In the production plants of Naber Konfeksion sh.p.k two lines used to train new employees for three months. Employees are trained on various processes like sewing, cutting, and buttoning. At the end of the training period, employees are evaluated from their supervisors and depending on the skills they have acquired are assigned to respective departments and divisions. Employees in the other two production plants, where most articles are made of delicate fabric, need one year to catch the same productivity as an experienced employee.[17] On the other hand, employees in Naber Konfeksion 2 sh.p.k and Grori Konfeksion sh.p.k require three months training to have the same productivity as an experienced worker, as the tasks performed are specialized and repetitive.

Engineers working in the preparation of models and the plotter department undergo a more professional training mostly in the head office. They are trained to master computer programs used in daily operations of the subsidiaries. Engineers follow also online trainings with their counterparts in the head office. These trainings are oriented towards preparation of samples of articles required before customers place their orders with Naber Moden.

Employees working in the finance department are trained to properly use accounting programs required to record financial transactions of the subsidiaries in Albania and to determine the monthly salary of each employee depending on the minutes required to complete the number of articles produced daily.

To continue, Naber Konfeksion sh.p.k is one of the clothing manufacturing enterprises in the country that together with the head office has prepared an educational program for local technical schools. The current, educational program for technical schools is outdated and not in line with the developments of the clothing industry in Albania. The management of Naber Konfeksion sh.p.k together with leading German experts in the clothing industry has drafted curricula to prepare students according to the needs of the market and to increase the availability of properly educated employees that are nowadays absent in the local labor market. This initiative has no cost for the Albanian Government as its implementation is a donation of Naber Moden.

Working hours in Naber Konfeksion sh.p.k are from Monday to Saturday between 7:00 am and 15:30 p.m. including a 30-min lunch break. The lunch break varies from one department to the other. Employees are required to use the personal identification

---

[17]Interview in June 2015 in the premises of Naber Konfeksion sh.p.k.

Table 6.23 Compensation of employees working in Naber Moden plant (2015)

No.	Category of employee	Monthly wage range
1.	Administration	570–700 EUR
2.	Head of Department/Division	400–450 EUR
3.	Blue collar employees	200–350 EUR depending on their daily production

Source Based on the information provided by Naber Konfeksion 1 sh.p.k

card to enter and to leave the company in order to record their working hours. Weekly working hours are in full compliance with the national Labor Code, with a maximum of 50 h per week. If employees work during national holidays the day off can be replaced after prior approval from the supervisor or can be added to the annual leave of employees. The annual leave for each employee is 28 working days including Saturdays and Sundays. All employees have 14 days of annual leave on August when production is interrupted while the rest of the annual leave can be taken based on the needs of each employee.

All employees have working uniforms assigned based on the responsibilities and tasks of employees. Naber Konfeksion sh.p.k offers daily transportation for the employees. During peak production seasons, temporal workers are hired.

Furthermore, employees are paid depending on the amount of articles they complete by working for a total of 480 min every day. For example, if the estimated time of making a collar is 1 min and the employee makes 600 collars in a day then he gets paid for 600 min and not for 480 (see Table 6.23). This system not only encourages employees to work more but also stimulates them to become more efficient and productive. Employees are paid 25% more if they work after 15:30 and are paid 125% if they work during holidays. All employees are insured. Representatives of trade unions participate in the meetings and are involved in drafting the internal regulation.

6.4 Valcuvia Alba sh.p.k

This section presents the third case study of the research. It is dedicated to the operational and production activity of Valcuvia s.r.l an Italian enterprise and its subsidiary Valcuvia Alba sh.p.k. Presentation of the case study starts with the activity of the group and it continues with the operational activities occurring in the subsidiary.

6.4.1 Overview of the Region of Berat

The region of Berat is situated in the south of Albania (see Fig. 6.20). The population of the region in 2017 was 131,942 inhabitants. The main town of the region, the city

Fig. 6.20 The map of Albania with the region of Berat circled. *Source* www.europe-atlas.com

of Berat is one of the oldest cities of Albania and has a unique cultural heritage and it is an important touristic center of Albania.

The processing industry in the region focuses on making the best of the resources available in the region. There is a growing stone processing industry, which turns natural stone into decorative ones used in the construction industry. Expansion of stone processing industry is replacing to some extend the use of imports in the local market. Handcrafting (embroidery, painting, iconography, etc.) is another area of the regional economy that is flourishing due to the high demand from foreign visitors. Agriculture remains the main economic activity of the region. The region is well-known for producing olives, figs, and wine. These products are exported in regional and European countries.

The clothing industry is an important industry for the regions. Before the 1990s, in the region of Berat was operating the state owned textile manufacturing enterprises that stopped production in the early 1990s. In the surroundings of the region there is a concentration of clothing manufacturing enterprises opened in the early 1990s.

In 2017, the GDP in current prices for the region of Berat amounted to 436 million EUR having a share of 3.77% in the country's GDP. In the same year, the GDP per capita for the region was 3,364 EUR at market prices (see Table 6.24). Agriculture has the main share in the region's GDP.

According to the National Statistical institute in 2018 in the region were 9,035 active enterprises, 5.54% of the total active enterprises in the country (see Fig. 6.21). Of the active enterprises in the region 1,631 are owned by females while only 12 of active enterprises have foreign ownership. Only 20 enterprises are jointly owned by foreigners and with Italian enterprises amounting to 17. The region is dominated by SMEs. In total there are 8,609 SMEs in the region that employ up to four persons.

Table 6.24 Table presenting the region of Berat (2017)

Indicator (2017)	Region of Berat	Albania
Population	131,942	2,876,591
Total GDP in (Mln/EUR)	436	11,564
GDP per capita in (EUR)	3,364	4,024
GDP composition in (%) of the total GDP		
Agriculture, forestry, and fishery (%)	49.60	22.70
Industry (%)	12.90	13.90
Construction (%)	7.90	10.20
Trade, transport, hotels, and restaurants (%)	10.10	18.10
Information and communication (%)	2.10	3.50
Financial and insurance activities (%)	1.40	2.80
Real estate activities (%)	3.10	6.60
Scientific, professional, administrative and supporting activities (%)	1.50	6.70
Public administration, education, health, and social activities (%)	10.20	12.50
Artistic, entertaining, and related activities (%)	1.40	3.00
Number of enterprises	8,763	162,452
Number of foreign owned enterprises	19	3,704

Source INSTAT, Chamber of Façon in Albania, Municipality of the city of Berat

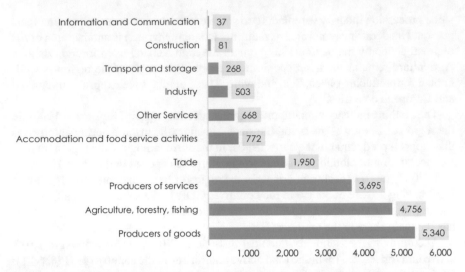

Fig. 6.21 Number of enterprises in the region of Berat (2018). *Source* National Statistical Institute, Business Register 2018

Enterprises employing up to 49 persons are 166 while those having over 50 employees are only 50.

6.4.2   The Valcuvia s.r.l Group

This section introduces Valcuvia s.r.l including its history and operating activity. It also covers the types of articles manufactured, the customers, suppliers, and the organizational structure in the head office of the enterprise.

Valcuvia s.r.l is an Italian enterprise specialized in the design and supply of underwear for women, men, and children. Its finished articles range from underwear for children to corsetry. As of 2015, the enterprise has the head office in Italy while production occurs in its subsidiary Valcuvia Alba sh.p.k located in Albania and subcontractors in Serbia and North Africa.

Valcuvia s.r.l is an Italian enterprise with over 30 years of experience in manufacturing of underwear (see Fig. 6.22). The head office of Valcuvia s.r.l is located in the north of Italy, in Cuveglio, province of Varese, and 80 km distant from Milan.

Since the establishment, Valcuvia s.r.l offers to its international customers a complete set of services ranging from research and development of raw materials to delivery of finished products ready to be sold individually. Valcuvia s.r.l operates with its customers in developing new articles by focusing on style, design, and preferences of customers. Valcuvia s.r.l provides to its customers the collection developed by the designers of the enterprise. This collection is updated each season with hundreds of new designs.

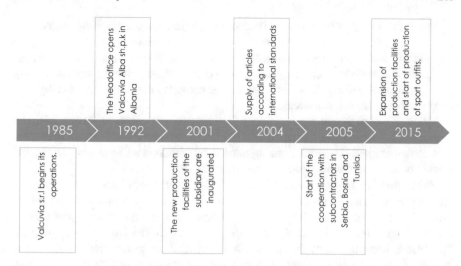

Fig. 6.22 Main events in the history of Valcuvia s.r.l. *Source* Based on the information provided by Valcuvia s.r.l

Valcuvia s.r.l is present for more than 15 years in Albania, with a manufacturing subsidiary. In addition, Valcuvia s.r.l relies also on subcontractors located in Serbia, Bosnia, and Tunisia. These subcontractors are monitored periodically in order to ensure that they manufacture articles according to the requirements of the customers. Subcontractors engage in manufacturing only during seasonal peaks when the Albanian subsidiary works under full capacity and is unable to accommodate additional orders.

In 2013, the total volume of sales was 4.8 million EUR while in 2014 total sales amounted to 5.3 million EUR experiencing an increase of 10.42% despite of difficulties in the European economy.

Valcuvia s.r.l is characterized by a high degree of market adaptability. Initially it was present only in Italy and later it entered the British market manufacturing for Arcadia group especially for Debenhams. Afterwards, it entered into the German, French, USA, Eastern Europe, and Latin America markets. For Valcuvia s.r.l it has not been easy to adjust to the requirements of various clients operating in markets with diverse customer taste and preference.

Since 2004, Valcuvia s.r.l is able to supply its products with the safety of the Oeko-Tex 100 Standard. This standard ensures that textile used in manufacturing of underwear does not contain or release substances harmful to human health.

Valcuvia s.r.l is also certified to ISO 9001:2000 for Quality Management. The ISO 9001:2000 is the new reference, recognized worldwide for the certification of quality management system of organizations in all industries and of all sizes. The 2000 revision of ISO 9000 (the third since 1987) has as its main objective the applicability to any type of business.

The ISO 9001: 2000 focuses on: (i) implementation of a management system, (ii) customer and his satisfaction, (iii) vision of the company as a set of processes in close relationship with each other aiming to provide products that consistently meet the standards of each customer. For Valcuvia s.r.l quality management means achieving the effectiveness and efficiency of its processes through knowledge, monitoring, and involvement of human resources.

Valcuvia s.r.l produces for a number of international brands (see Table 6.25). The number of customers has increased and has become more diverse within the years. The customers sell the articles manufactured by Valcuvia s.r.l not only in the EU market but also in Russia, China, Latin America, etc.

Production is based on imported raw materials. Local suppliers of raw materials are not present as they do not meet the standards imposed by international customers. Inputs like card boxes and labels are also imported as each customer provides its own packaging inputs. Packaging inputs are imported mostly from Italy while most of fabric is imported from Turkey or China (depending on the preferences of the customer). Imports from Asian countries include a variety of fabrics like cotton, lace, elastic etc. Yearly, Valcuvia s.r.l imports over 50 tons of fabrics to meet its annual production. Due to its presence in diverse markets and regions, Valcuvia s.r.l has worked with a considerable number of suppliers.

A good cooperation between the enterprise and suppliers is based on a clear division of tasks between the head office and the subsidiary. The head office discusses with customers on the inputs required to meet orders. They establish direct contacts with suppliers of raw materials and perform rigid quality control tests on samples of raw materials including endurance, stretching, thermal resistance, color fading, etc. in the quality control laboratory of the head office. If tests are satisfactory the sample of raw materials are sent to the subsidiary in Albania. In Valcuvia Alba sh.p.k samples undergo a second quality control on possible marks, holes, missing threads, etc. from the responsible unit and only after the sample undergoes a final test.

Table 6.25 Main customers of Valcuvia s.r.l

No.	Main clients	Products for each client	Output in % for 2015
1.	Guess	Slip, boxer, shirts, brass, tops, shirts	20
2.	Versace	Slip, boxer, shirts, brass, tops, shirts	5
3.	Schiesser	Slip, boxer, brass, children, camisole	10
4.	Disney	Slip, boxer, brass	3
5.	Santini	Sport line (shirts)	10
6.	Go Sport	Slip, boxer, brass for sports	3
7.	Pier Roberti	Slip, boxer, shirts, brass, tops, shirts	15
8.	Womo	slip, boxer, shirts, tops	16
9.	Esselunga	Slip, boxer, shirts, brass, tops, shirts, pyjamas	10
10.	Myllen	Slip, boxer, shirts, brass, tops, shirts, camisole	8

Source Based on the information provided by Valcuvia Alba sh.p.k

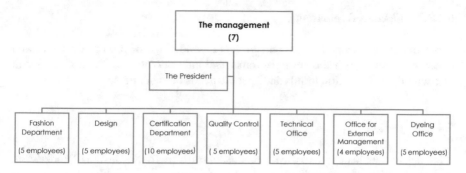

Fig. 6.23 Organizational structure of in Valcuvia s.r.l. *Source* Author representation based on the information provided by Valcuvia s.r.l

The annual rejection rate in the total quality of imported raw materials is 2–3%. Meetings with suppliers are very frequent and they reach up to 12 times in a year. Meetings serve mainly: (i) to suggest technical improvements on existing materials, (ii) to introduce seasonal fabrics especially for limited editions, (iii) to keep suppliers updated on the trends of clothing industry, (iv) to perform a cross-checking data analysis that serves to pinpoint new areas of cooperation with existing pool of suppliers. The management of Valcuvia Alba sh.p.k states that a close and efficient cooperation with suppliers is a key factor in achieving quality within short delivery times.

Oriented towards efficiency Valcuvia s.r.l has established a delivery system for its finished garments. For the transportation, a company is subcontracted. Trucks are used to deliver finished article to the head office in the north of Italy. Albanian seaports are used to cross the Adriatic Sea and reach the port of Bari in the south of Italy and to continue afterward in the Italian national road transportation network to reach the final destination. Valcuvia s.r.l is not responsible to deliver goods to every customer. In the head office finished articles are placed in the vast storage houses monitored through an inventory system which records the inflow/outflows goods and shows the stock at any point of time. When orders are placed in the storage unit customers are informed to take them out.

As of 2015 the head office of Valcuvia s.r.l is still located at its founding location and continues to be a family run business (see Fig. 6.23). The company has 40 employees in the head office and 145 in the subsidiary in Albania.

6.4.2.1 The Management

It includes stakeholders and the senior management (finance, operations, strategy etc.) that oversees operations in the head office and in the subsidiary of the enterprise. It is the main decision making body on the operations of Valcuvia s.r.l and in charge of the company's expansion strategies including establishments of new manufacturing sites, promotion activities, and entering in new markets.

6.4.2.2 Fashion Department

It is a department responsible to ensure that a wide range of products are renewed every six months with the spring/summer and autumn/winter collection totally in line with the new fashion trends and specific needs of customers.

6.4.2.3 Design

It is a department in charge to use sophisticated software for the design of underwear. It is also able to fit the needs of customers into easy to use graphics that can enter into production in the subsidiary.

6.4.2.4 Certification Laboratory

It is a department that assures the quality of raw materials for the manufacturing of samples and the dry cleaning for small batches. The certification laboratory operates in close cooperation with the quality control department.

6.4.2.5 Quality Control

It is a department that performs ongoing controls during the production cycle and the final check on the manufactured products before they are shipped to corresponding customers. The laboratory is capable of performing the required tests mainly in the textile and clothing industry, using both international regulations an those tested and evaluated internally by Valcuvia s.r.l. Since 2000, the enterprise is certified and audited annually by Arcadia Group Limited.

6.4.2.6 Technical Office

It is a department that gets constantly informed on recent innovations that are introduced in clothing manufacturing. This department guides the head office to apply this changes both in terms of design and subsequently in production. Valcuvia s.r.l was in the past one of the first companies to adopt automatic cutting, CAD, laser cutting, and ultrasound stitching.

6.4.2.7 Office for External Management

It is a department responsible to monitor daily production in the subsidiary and to maintain continuous communication and cooperation with customers and the selling points.

6.4.2.8 Dyeing Office

It is a department in charge to realize in a short time any dyeing of samples including fabrics, lycra, and lace in every shade of color. The office is equipped with the latest technology that can give the maximum precision in dosage in the repetition of the same colors. It uses machines for continuous dyeing of tapes, pressure, flow (for dyeing fabrics) and tumblers (fabrics, hooks etc.).This allows the enterprise to shorten the time needed in the dyeing process, to achieve the highest accuracy in color reproduction, and to yield maximal strength. With the same machines, the enterprise is also able to dye fabrics required for producing small quantities.

Referring to value adding activities of clothing manufactures based on those presented in Chap. 3, the figure below present the value adding activities that occur at Valcuvia s.r.l (see Fig. 6.24).

As seen, the activities having the higher gains occur in the head office, leaving only production to occur in its subsidiary in Albania. The president of Valcuvia s.r.l closely monitors production in the subsidiary and decides on the expansion of production including new lines of sport outfits and new operating facilities.

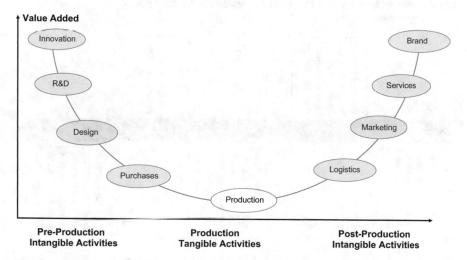

Fig. 6.24 Value adding activities of Valcuvia s.r.l. *Source* Based on the information provided by Valcuvia s.r.l

6.4.3   The Subsidiary Valcuvia Alba sh.p.k the Clothing Manufacturing Subsidiary

This section presents the operational activity of Valcuvia Alba sh.p.k, which reflects the information obtained during fieldwork in the premises of the subsidiary.

In 1992, immediately after the fall of the socialist regime, Valcuvia s.r.l opened its subsidiary in Albania (see Fig. 6.25). However, the subsidiary became a fully operating enterprise only in December 1994, after the local private sector started to function under the market economy. In 1992, the subsidiary had only four employees working with five sewing machines in the basement of a house set up before the 1990s. Despite of the difficulties the four employees together with the ongoing support of head office in Italy managed to expand the activity and in 2001 the subsidiary constructed its own facilities occupying an area of 8,000 m^2.

The management of Valcuvia s.r.l decided to open a subsidiary in Albania mainly because of low labor cost and the proximity the country has to the Italian and European market. However, it is one of the few foreign and Italian enterprise that is located in the region of Berat.

The main reason for the management was to benefit from the skills that the population has gained from textile manufacturing that was the heart of the city's economy during the socialist regime. Textile manufacturing stopped after the 1990s and the majority of workers were unemployed. It was very advantageous for Valcuvia s.r.l to train local employees upon the skills they have obtained from working in textile manufacturing during the socialist regime.

Referring to value adding activities of clothing manufactures Fig. 6.26 presents the value adding activities that occur at Valcuvia Alba sh.p.k.

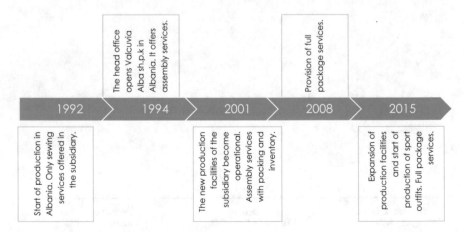

Fig. 6.25 Main events in the history of Valcuvia Alba sh.p.k. *Source* Based on the information provided by Valcuva Alba sh.p.k

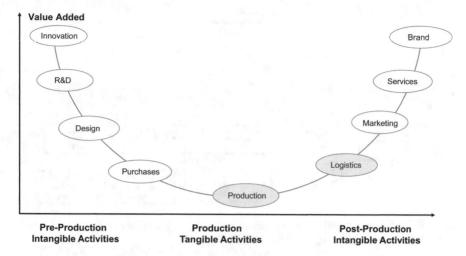

Fig. 6.26 Value added activities in Valcuvia Alba sh.p.k. *Source* Based on the information provided by Valcuvia Alba sh.p.k

Initially the subsidiary provided only sewing. Under direct supervision from the head office more processes and production lines were added in its facilities in Albania (see Fig. 6.26).

The subsidiary is lead by two Italian administrators that are appointed from the head office. They are in charge for overall operations of the subsidiary ranging from acceptance of raw materials to the delivery of finished articles to the head office. Administrators have deep understanding of clothing manufacturing and on the operations of the industry in Albania. Administrators are able to provide on the spot technical support to employees working in production lines and in the quality control department. The administrators reside in Albania on the working days and during weekends, they go to Italy. Currently, the organizational structure of Valcuvia Alba sh.p.k is presented in Fig. 6.27:

In running the subsidiary, administrators are assisted by the general director who oversees overall production including communication with local staff, relations with local agents including tax authorities, custom authorities, and approval of training programs for new employees. The general director has been working for Valcuvia Alba sh.p.k for more than 10 years and has strongly supported its expansion in the city of Berat. A secretary supports the administrators and the general director in their duties, including arrangements of various meetings, scheduling of events, travel arrangements, and management of office supplies.

Responsible for financial transactions is the accountant of the subsidiary. The accountant administrates the online national system for tax declarations, payment of employees' social contributions, customs declarations on imports of raw materials and the level of exports. The responsible records the presence twice a day and records on the absences, permissions or sick leaves for the each employee. The monthly presence is also included in the overall evaluation of employees.

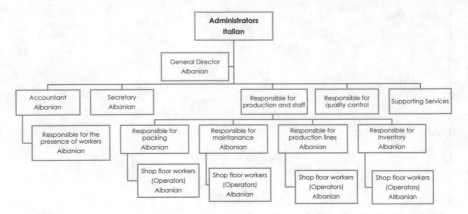

Fig. 6.27 The organizational structure of Valcuvia Alba sh.p.k. *Source* Author representation based on the information provided by Valcuvia Alba sh.p.k

In Valcuvia Alba sh.p.k the responsible for production and staff is the one that daily monitors all the three production lines. He closely cooperates with employees in charge of each production line. The responsible provides solutions to problems encountered during production, assigns the level of daily production to be manufactured, introduces new models to each production line, and ensures that samples meet all the requirements demanded by each customer. As Valcuvia s.r.l relies on division of labor it has assigned one responsible for each production line. Employees in charge for each production line strictly monitor completion of the steps required for a particular article to go through the line.

In the subsidiary, there is a unit responsible for inventory that is divided in two storage places. In the first one are stored raw materials needed in production lines while the second one stores the finished articles that are ready to be transported to the head office or directly to customers. The main objective of the inventory unit is to keep detailed records of raw materials that enter in the subsidiary and the finished goods that are exported.

Before all the merchandise that enters or exits the premises of the subsidiary, the inventory unit performs rigid quality control checks. After finished articles undergo quality control checks they pass through the packing unit where goods are packed based on the requirements (boxes, labels, folding, numbering) of each customer.

Finally, the subsidiary has a unit of supporting services that monitors the power supply, ventilation and the alarm system are properly functioning so that the manufacturing of articles is not interrupted. Moreover, in this category are included employees cleaning the working space from materials remaining after cutting and sewing.

In 2001, the enterprise settled in an area of 8,500 m^2 half of which were occupied by manufacturing facilities. The investment made in the new manufacturing facilities constituted at that time the largest foreign investment made in the country. On its

Table 6.26 Investments in Valcuvia Alba sh.p.k (2015)

Category of investment	Amount in EUR
Investments in operating facilities	
Building	1,000,000
Ventilation	50,000
Electricity generators	600,000
Guest house/canteen	500,000
Security	60,000
Investments in machines	
Cutting machines	70,000
Sewing machines	1,000,000
Printing machines	150,000
Molding machines	100,000
Ultrasound machines	80,000
Pressing machines	100,000
Investments in information technology	
Hardware	35,000
Software	15,000
Total of investments	3,810,000

Source Based on the information provided by Valcuvia Alba sh.p.k

opening day, the Albanian president inaugurated the facilities. Even today, Valcuvia Alba sh.p.k continues to be the major foreign investment in the region of Berat.

The total investment for the operating facilities exceeds 1.5 million EUR (see Table 6.26). In the main building are placed production lines, administration, the packing, and the quality control department of the subsidiary. In the second building is placed the inventory of the subsidiary, and in the third one that is undergoing renovations a new production line for sports outfits will be located.

The subsidiary has given special attention to ensuring uninterrupted power supply at all times. Inadequate power supply is cited as one of the major problems encountered in the production process. The investment made on the power supply generator ensures continuity of production. The enterprise has also invested in the canteen that may serve up to 100 employees during lunch breaks and in five apartments that host the administration and the guests of the subsidiary including client representatives of customers and suppliers of raw materials.

Investments in machines include: laser cutting machines, machines to prepare all sizes of bras cups, machines for printing the designs, machines to eliminate edges through heating and seam, automatic machines for application of rhinestones, and machines capable of cutting and fixing cold materials using ultrasound. Machines are of Italian and Japanese brands including Mecasonic and Juki. With exception of few machines that are used in the training of new employees, the rest are programmed to perform various processes depending on the type of fabric and stitches (single thread, double needle, zigzag etc.). In addition, the subsidiary has three cutting machines used mostly to cut cotton slips, shirts, and pyjamas that operate with ultrasound technology.

Investments were made also on machines used to prepare different bras cups through heating and high pressure. Such machines are capable to prepare any cup on any fabric that is preferred by the customer. During peak seasons, additional or specific machines are required to meet the temporary needs of a customer. The head office supplies them as in Italy there is a large inventory of modern machines that is regularly updated to keep up with market trends and demands.

In the subsidiary investments, are made also in information technology. Administration has modern computers supported with advanced servers and software used to monitor the level of inventory in the subsidiary including imports of raw materials and exports of finished articles. In the same time, the subsidiary has purchased modern programs for sample cutting and development of new models.

Production starts after head office sends to Valcuvia Alba sh.p.k the raw materials together with the corresponding model in order to prepare the samples that need to be controlled by customers before they determine the quantity and the delivery date of their order (see Fig. 6.28).

Samples are prepared depending on the sizes asked by the customer. Usually, samples are prepared in small, medium, and large sizes. However, different customers may ask for special sizes like five extra large or extra small (see Table 6.27).

Valcuvia Alba sh.p.k adjusted part of its facilities to manufacture sport outfits especially those used in cycling including quality outfits for international cycling competition like Giro d'Italia and Tour de France. These special outfits are expected to be manufactured by using special fabrics including silver ionized fabrics that ensure adequate perspiration in order to maintain an optimal level of body temperature. Valcuvia Alba sh.p.k is preparing also a training program for employees that will be involved in the production of cycling outfits.

The other project, which Valcuvia Alba sh.p.k is expecting to introduce is with Schiesser, the German underwear brand. In this project Valcuvia Alba sh.p.k will be

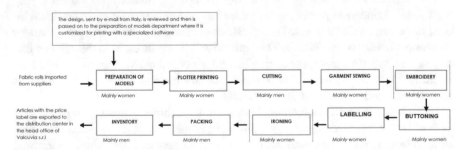

Fig. 6.28 Production flow (Embroidery and ironing are part of the production cycle only on specific articles. The vertical lines indicate the production flow of articles that are not subject to embroidery and ironing) in Valcuvia Alba sh.p.k. *Source* Author representation based on information provided from Valcuvia Alba sh.p.k

Table 6.27 Output of Valcuvia Alba sh.p.k (2010–2015)

No.	Item	Output (pairs) 2010	Output (pairs) 2015	Change in output (%)
1.	Slip	100,000	160,000	60.00
2.	Boxer	220,000	260,000	18.18
3.	Brass	98,000	120,000	22.45
4.	T-Shirts	18,000	1,150,000	6,288.89
5.	Pyjamas	2,000	30,000	1,400.00
6.	Tops	210,000	220,000	4.76
7.	Camises	7,000	20,000	185.71
8.	Total	655,000	1,960,000	199.24

Source Based on the information provided by Valcuvia Alba sh.p.k

responsible to produce 7 different types of female bras that will be tailored to the female anatomy meaning that for each bra cup size the waist and the strips attached to the bras will be adjusted to cup size.

After setting into new operating facilities Valcuvia Alba sh.p.k tried to cooperate with different subcontractors located in the city of Vlora, where it identified local enterprises available at the shortest distance. However, this cooperation was for less than two years as subcontractors could not meet the quality standards agreed by the head office and their customers. Having trouble in finding appropriate subcontractors in Albania induced the head office to seek alternative companies offering these services in Serbia, Bosnia Herzegovina, and Tunisia. The management of Valcuvia Alba sh.p.k emphasized that despite of the expansion in their production they are not looking for new subcontractors in Albania.

As of 2015, Valcuvia Alba sh.p.k has 145 employees out of which 16 are male and 124 are female (see Table 6.28). Female employees dominate the three production lines, the accounting department, and in tracking the daily presence of employees. However, in the subsidiary there are two male employees working with sewing machines. In peak seasons, especially between April and August the subsidiary hires between 35 and 40 seasonal employees. Overall, male dominate in management positions with 10 employees while females are only six.

Among employees are also those involved in the maintenance including mechanics, drivers, guards, gardeners, and cleaners. Mechanics assist in the daily operations

Table 6.28 Level of employment in Valcuvia Alba sh.p.k (1992–2015)

No.	Year	Level of employment
1.	1992	4
2.	2001	100
3.	2012	120
4.	2015	145

Source Based on the information provided by Valcuvia Alba sh.p.k

of the firm. Each of them has over 20 years of experience in maintaining machines and are able to fix on the spot the majority of problems occurring during production. In the same time, they are capable to correctly identify the parts needed to be replaced without causing long interruptions in production. In addition, cleaners are essential in keeping a tidy working environment especially in removing the waste generated after cutting and sewing.

Italian administrators focus on: (i) suggesting new techniques for increasing labor productivity, (ii) participating in administrative meetings aimed at monitoring operations in the subsidiary, (iii) setting future objectives, (iv) informing the head office on the level of operations and performance of the subsidiary, and on (v) providing solutions to difficulties in production that cannot be solved by local administration.

Criteria for selection of employees depends on the degree of responsibility and the difficulty of assigned tasks. Administrators are directly hired and appointed by the head office. They have a university degree speak at least two foreign languages, and possess deep knowledge on clothing manufacturing. The general director is selected by the administrator of Valcuvia Alba sh.p.k and is approved by the head office. The general director possesses not only extensive knowledge on clothing manufacturing but also a good understanding of the business environment in the country.

Division heads are selected from administration based on previous experiences in the industry. They need to have a secondary or higher degree, speak at least one foreign language, and possess basic level of computer capabilities. In total, there are seven division heads in Valcuvia Alba sh.p.k. Employees have the opportunity to be promoted. For example, after several years employees that start working in sewing lines can be become a division or a department head. When it comes to hiring blue-collar employees for processes like cutting, sewing, and packing the management of the enterprise has difficulties in finding quality employees willing to learn new processes and techniques. New employees are found through current employees or advertising in nearby surroundings of the subsidiary. Support from regional employment centers has been very limited in finding female employees.

Evaluation of employees is critical for Valcuvia Alba sh.p.k as it is directly related to improving the quality of finished articles. Evaluation criteria varies by position, responsibilities, and the tasks required to perform. For low to medium level employees, evaluation criteria are the number of goods completed daily, commitment, teamwork, precision, and amount of overtime hours. Head of departments and division are evaluated based on the ability to instruct and guide their unit in meeting targets, reliability, and improved performance. The evaluation form is prepared in the head office and approved by the management of its subsidiary in Albania. The form is structured in such a way to ensure not only inclusion of production cycle but also to allow quality cross-checking on a daily basis.

Valcuvia Alba sh.p.k provides training programs for its employees. Trainings are organized for all level of employees. In the subsidiary there is a special line to train new employees for 60 days. Employees are trained on various processes like sewing, cutting, buttoning. At the end of the training period employees are evaluated from their supervisors and depending on the performance they are assigned to respective divisions.

Table 6.29 Compensation of employees in Valcuvia Alba sh.p.k (2015)

No.	Category of employee	Monthly wage range
1.	Administration	700 EUR
2.	Head of department/division	350–400 EUR
3.	Blue collar employees	200–300 EUR depending on the quantity of production

Source Based on the information provided by Valcuvia Alba sh.p.k

Employees responsible for various units have undergone training in the head office in Italy. In total eight employees have had international training. These employees are considered as assets of the enterprise and in several cases they are the ones to suggest possible solutions to the head office. Employees responsible for finances are trained in two directions. Firstly, they are trained for accounting programs required to record financial transactions of the subsidiary in Albania. Secondly, they learn how to determine the monthly salary of each employee depending on the number of goods produced on a daily basis. Trainings in Valcuvia Alba sh.p.k include also topics like safety and health in the workplace, maintenance of sewing machines, training and basic knowledge of machines, training against discrimination of color, religion, and origin.

Working hours at Valcuvia Alba sh.p.k are from Monday to Saturday between 8:00 am and 16.30 p.m. including a 30-min lunch break. Employees are required to use the personal identification card to enter and to leave the company in order to record their working hours. Weekly working hours are in full compliance with the national Labor Code, with a maximum of 50 h per week. Overtime hours are paid 25% more and working hours during national holidays are paid 125%. National holidays can be replaced another day after approval with the supervisor or can be added to the annual leave of employees.

The annual leave for each employee is 28 working days including Saturdays and Sundays. All employees have 14 days of annual leave on August when production is interrupted and the rest of annual leave can be taken based on the needs of each employee. All employees are required to follow the set of safety and emergency rules and they all have their special uniform customized based on responsibilities and tasks of each employee. The subsidiary covers daily transportation for all its employees. Staff turnover is very low at 1% annually. Employees are paid depending on the quality and quantity of articles manufactured (see Table 6.29). All employees are insured and are paid in a timely manner without delays. Representatives of trade unions participate in the meetings and are involved in drafting of new staff regulation.

6.5 Industria Ballkanike sh.p.k

This section presents the fourth case study of the research. It is dedicated to the operational and production activity of the Greek enterprise Industria Ballkanike and its subsidiary Industria Ballkanike sh.p.k. The subsidiary has two production plants. Presentation of the case starts with the activity of the group and it continues with the activities occurring in the subsidiary.

6.5.1   Overview of the Region of Korca

The region of Korca is situated in the southeastern part of Albania (see Fig. 6.29). The region of Korca is very mountainous with an average height of 850 m on the sea level. In 2017, the region had a population of 214,321 inhabitants. Korca has a favorable geographical position because it is situated between trading roads that join Albania with Macedonia (47 km) and Greece (35 km). The food industry is very developed and focused on dairy and honey products. Artisanship in the region has ancient traditions. It is distinguished for the processing of wool, stone, wood, metal, etc. Tourism has a potential to attract foreign visitors due to the presence of natural attractions and cultural monuments. The majority of the enterprises in the clothing industry are in the form of Greek-Albanian joint venture operating with imported raw materials and exporting of goods overseas.

In 2017, the GDP in current prices for the region of Korca amounted to 637 million EUR, contributing to the Albanian GDP with 5.51% (see Table 6.30). In 2013, the GDP per capita in the region of Korca was 3,001 EUR Agriculture remains the main economic activity of the region contributing to the GDP of the region by 41.10% in 2017. The region is well-known for producing apples, beetroots, and beans. These products are exported to neighboring and European countries.

According to the National Statistical institute, in 2018, in the region of Korca were 14,035 active enterprises, 8.61% of the total active enterprises in the country (see Fig. 6.30). Of the active enterprises in the region 2,559 are owned by females while only 66 out enterprises have foreign ownership and 61 have joint ownership between foreigners and local citizens. The region is dominated by small enterprises in total 13,261 that employ up to four persons. Enterprises employing up to 49 persons are 323 while those having over 50 employees are only 82.

6.5.2   Industria Ballkanike Group

This section introduces the head office of Industria Ballkanike including its history and operating activity. This section includes also the type of articles manufactured, the customers, suppliers and the organizational structure of the head office.

The group of Industria Ballkanike consists of a head office dealing mostly with administrative issues and one subsidiary Industria Ballkanike sh.p.k located in Albania. The subsidiary is composed of two production plants Industria Ballkanike 1 sh.p.k and Industria Ballkanike 2 sh.p.k.

Fig. 6.29 The map of Albania with the region of Korca circled. *Source* www.europe-atlas.com

Industria Ballkanike is a clothing manufacturing enterprise founded in 1992 in the city of Thessalonica, Greece (see Fig. 6.31). It started to operate in the region of Korca in 1993 being one of the first Greek enterprises that located its production in Albania. The enterprise is the initiative of two brothers that after the socialist regime was abolished believed in the potential of economic development of Albania. The

Table 6.30 Table presenting the region of Korca (2017)

Indicator	Region of Korca	Albania
Population	214,321	2,876,591
Total GDP in (Mln/EUR)	637	11,564
GDP per capita in (EUR)	3,001	4,024
GDP composition in (%) of the total GDP		
Agriculture, forestry, and fishery (%)	41.10	22.70
Industry (%)	9.20	13.90
Construction (%)	7.50	10.20
Trade, transport, hotels, and restaurants (%)	16.00	18.10
Information and communication (%)	2.00	3.50
Financial and insurance activities (%)	2.60	2.80
Real estate activities (%)	5.00	6.60
Scientific, professional, administrative and supporting activities (%)	1.80	6.70
Public administration, education, health and social activities (%)	11.80	12.50
Artistic, entertaining, and related activities (%)	3.10	3.00
Number of enterprises	13,814	162,452
Number of foreign owned enterprises	61	3,704

Source INSTAT and the Chamber of Façon in Albania

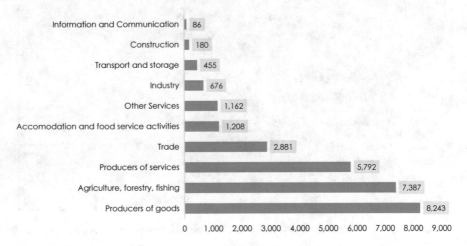

Fig. 6.30 Number of active enterprises in the region of Korca (2018). *Source* National Statistical Institute (INSTAT), Business Register 2018

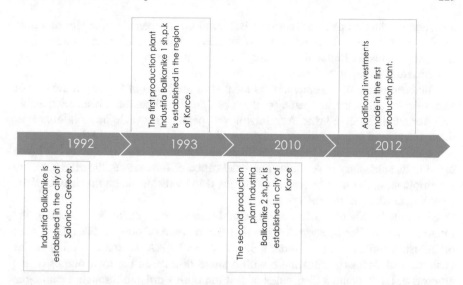

Fig. 6.31 Main events in the history of Industria Ballkanike. *Source* Based on the information provided by Industria Ballkanike sh.p.k

main reasons they decided to transfer production to Albania were: the (i) proximity with the Greek border, (ii) the cheap labor force, (iii) the tradition of the Korca region in the textile industry, (iv) the vicinity with the European market, (v) the similar social characteristics of the region with the Greek ones.

The enterprise continued its manufacturing activity even during the civil war in Albania in 1997. In its beginnings, the enterprise had only 10 employees and 15 machines. Its first customers were Greek brand names like BSB, CAM, and On Line. Gradually, the enterprise became a subcontractor of international brands like Decathlon, Assos, C&A etc.

Table 6.31 summarizes the main characteristics of production plants possessed by Industria Ballkanike in Albania. In the early 2000s, the management of the enterprise decided to concentrate only on the production of swimwear and sport outfits. In 2010,

Table 6.31 Production plants of Industria Ballkanike in Albania

Plants of industria Ballkanike in Albania		
Indicator	Industria Ballkanike 1	Industria Ballkanike 2
Products	Bikini, swimsuits, shorts, shirts, children line	Bikini, swimsuits, shorts, shirts, children line
Jobs (June, 2015)	391	138
Industrial facilities m² (June, 2015)	4,000	1,500
Date of setting up	1993	2010

Source Based on the information provided by Industria Ballkanike

a second production plant of Industria Ballkanike was opened in the city of Korca through acquisition of the previously state owned clothing manufacturing enterprises. Currently, Industria Ballkanike is considered one the main employers in the region of Korca amounting to 529 employees.

The head office of the enterprise is located in the North of Greece in the city of Thessalonica and since the opening of the subsidiary in Albania have been responsible only for administrative tasks. Administrative operations include financial management, tracking of orders, customer relations, and monitoring activity of the subsidiary. Since the establishment of Industria Ballkanike, the manufacturing activity has taken place in its subsidiary in Albania. The head office of Industria Ballkanike has only six employees. The team in the head office is lead by the president that is one of the founding brothers of the enterprise.

Since the 2000s the main customer of Industria Ballkanike is Decathlon, the international retailer of sports outfits that has a share of 60% in the total output of Industria Ballkanike (see Table 6.32). Asos and C&A are the two other major customers of Industria Ballkanike with a share of 25% in the total output of the enterprise. La Redoute a distributor of fashion outfits ordered through a catalogue for home delivery has only a share of 5% in the total output. The category of goods manufactured for each customer include bikini, shorts, swimsuits, and shirts.

Industria Ballkanike receives in its subsidiary all raw materials required for clothing manufacturing. The management of the enterprise tried to cooperate with local suppliers that manufacture plastic bags but the customers found it to be very expensive and not in line with the quality demanded. Yearly, Industria Ballkanike imports around 15 tons of fabrics to meet its annual production. Considering that the main article manufactured in Industria Ballkanike is swimwear for international brands, the fabric is fully imported from Turkey and China. Raw materials reach the premises

Table 6.32 Main customers of industria Ballkanike and output in % (2015)

No.	Main customers	Products for each customer	Output in 2015 (%)
1.	Decathlon	Bikini, shorts, swimsuits, shirts	60
2.	C&A	Bikini, shorts, swimsuits, shirts	15
3.	Asos	Bikini, shorts, swimsuits, shirts	10
4.	Intersport	Bikini, shorts, swimsuits, shirts	10
5.	La Redoute	Bikini, shorts, swimsuits, shirts	5

Source Based on the information provided by Industria Ballkanike

of the subsidiary of Industria Ballkanike only after they are fully tested in the laboratories of the customer. The customers establish direct contacts with suppliers and perform rigid quality control tests on samples including endurance, stretching, thermal resistance, color fading etc. If tests are satisfactory, the samples of raw materials are sent from the head office to the subsidiary in Albania. In the subsidiary, samples of raw materials undergo a second quality control on possible marks, holes, missing threads, etc. After the second quality control raw materials are ready to go to production lines.

The annual rejection rate of raw materials is 2–3%. Industria Ballkanike has had over the years the same suppliers. Furthermore, inputs like card boxes and labels are imported as each client provides its own packaging inputs. Packaging inputs are imported mostly from Greece. Meetings of the management of Industria Ballkanike with suppliers of raw materials occur up to three times a year mostly when the management of Industria Ballkanike is invited by its customers in the meetings they have with the suppliers of raw materials. These meetings serve: (i) to suggest technical improvements on existing materials, (ii) to introduce seasonal fabrics especially for limited editions, (iii) to keep suppliers updated on the trends of the clothing industry, (iv) to perform a crosschecking data analysis, and (v) to pinpoint the new areas of cooperation with the existing pool of suppliers. The management of Industria Ballkanike states that a close and efficient cooperation with suppliers is a key factor in manufacturing quality products within the requirements of each customer.

Industria Ballkanike has established a delivery system for its finished articles to meet the short delivery times required by its customers. It has acquired both its own transporting vehicles and has subcontracted a delivery company. The vehicles owned by Industria Ballkanike are used to bring in the premises of the enterprise, goods between the two production plants in order to undergo all processes required for production. In the same time, these trucks are used to deliver finished articles to nearby ports of Albania through which they reach end customers. On the other hand, the vehicles of the subcontracted company are used to deliver finished articles to end customers like Decathlon, La Redoute, etc.

Figure 6.32 presents the value added activities of Industria Ballkanike according to value added activities presented in Chap. 3. Value added activities are limited to distribution to end customers and marketing of the services by the enterprise.

6.5.3   Subsidiary Industria Ballkanike sh.p.k the Clothing Manufacturing Enterprise

Manufacturing in Industria Ballkanike sh.p.k started with the opening of the first production plant in 1993 and continued with the opening of the second production plant in 2010 (see Fig. 6.33). As of 2012, the head office is undertaking additional investments to expand the existing production facilities to accommodate the increase in the orders from customers.

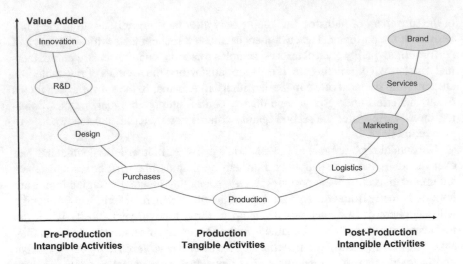

Fig. 6.32 Value added activities in Industria Ballkanike. *Source* Based on the information provided by Industria Ballkanike sh.p.k

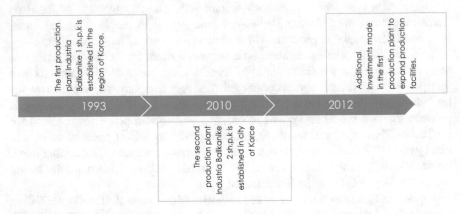

Fig. 6.33 Main events in the history of Industria Ballkanike sh.p.k. *Source* Based on the information provided by Industria Ballkanike sh.p.k

The level of employment in Industria Ballkanike has increased since 1993 in which the subsidiary had only 10 employees. As of 2015, Industria Ballkanike has 529 workers (see Table 6.33).

The increase in the level of employment generated by Industria Ballkanike results mainly for the increase in the operational activity and acquisition of additional operating facilities that resulted in the opening of the second plant of the subsidiary. Additional, employment is generated from expansion of existing production facilities and adding up of more cutting and sewing lines.

Figure 6.34 presents the value added activities of Industria Ballkanike sh.p.k

Table 6.33 Level of employment in Industria Ballkanike (1993–2015)

No.	Year	Level of employment
1.	1993	10
2.	2009	370
3.	2012	480
4.	2015	529

Source Based on the information provided by Industria Ballkanike

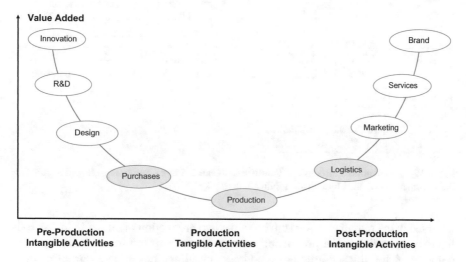

Fig. 6.34 Value added activities in Industria Ballkanike sh.p.k. *Source* Based on the information provided by Industria Ballkanike sh.p.k

according to value added activities presented in Chap. 3. Value added activities are limited to production.

The organizational structure of Industria Ballkanike sh.p.k is presented in Fig. 6.35. The president of Industria Ballkanike is in charge of setting the strategy and operations in the two subsidiaries. The president is the main decision making authority together with the administrative staff in the head office. The president is also the only foreign staff working in the two production plants.

In running the two production plants, the president is assisted by the General Administrator, a native of the region of Korca who is in charge of the operations of the two subsidiaries including communication with custom and tax officials. The general administrator proposes structural changes in the operations of the subsidiary after extensive consultations with production managers and the administration. Finally, the general administrator is responsible for hiring new employees in production. The general administrator supervises the manager for production and the administration in the two production plants.

The production manager is a woman with over twenty years of experience in the clothing manufacturing industry. The production manager supervises all units

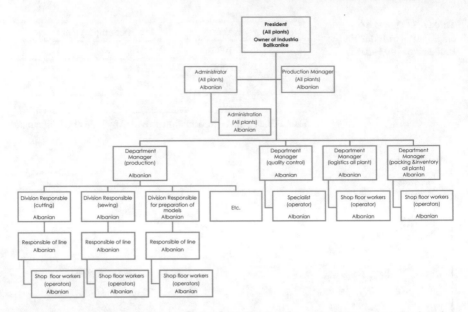

Fig. 6.35 Organizational Structure of Industria Ballkanike. *Source* Author representation based on the information provided by Industria Ballkanike sh.p.k

involved in the manufacturing activity in the two production plants of Industria Ballkanike. In addition, the production manager is responsible for ensuring on time delivery of the finished articles, to address any difficulties arising during production, and to administer the demands of employees assigned to various tasks during manufacturing. The key departments are:

6.5.3.1 Quality Control

It is a department that is responsible for controlling of raw materials used in manufacturing of goods. The quality control in Industria Ballkanike consists in physical control like identification of any holes, inappropriate stamping. Chemical, physical, and stretching tests of raw materials are performed in the premises of each customer before they are shipped to Industria Ballkanike to start production. The post-production quality control of raw materials consists in performing a final check on finished articles to ensure that they are fully in line with the requirements of the customers before they are send to the final destination.

6.5.3.2 Production

It is a department that covers several units. The units composing this department are cutting of raw materials, sewing, ironing, embroidery, cleaning of finished goods, and preparation of models for production. The second production plant is specialized in cutting and embroidery while the first production plant is focused on assembly services like sewing, limited cutting, and reparation of defects.

6.5.3.3 Packing and Inventory

It is a department in which finished articles are placed into appropriate boxes and where the quantity sent to each customer is recorded in the inventory system. In addition, this department is responsible to monitor the quantity of raw materials available in Industria Ballkanike sh.p.k and to report on a weekly basis to the production manager the quantity that can be used for production.

6.5.3.4 Logistics

It is a department responsible for the delivery of finished goods. Finished goods are delivered by subcontracting a local company that owns large trucks that periodically go to Greece while trucks owned by Industria Ballkanike go to the port of Durres where finished articles are shipped to the European market. The daily movement of goods from one production plant is done through the vans owned by Industria Ballkanike.

6.5.3.5 Administration

It is a department that depends on the general administrator. Administration in Industria Ballkanike includes the finance and accounting units that is responsible for monthly tax declaration, payables and receivables resulting from operations in Industria Ballkanike, and the monthly payrolls of the staff in two production plants. In addition, in the administration are included also the supporting services part of which are security guards, cleaners, drivers, the IT staff, and present during the production hours.

Investments made in Industria Ballkanike sh.p.k are easily spotted. The first production plant is located in the suburbs of the region of Korca while the second production plant is located in the city of Korca. The investment made in the first production plant Industria Ballkanike 1 sh.p.k covers a surface area of 4,000 sq/m. The investment was finalized in 2012. The amount invested in its facilities is over 900,000 EUR while investments in the second production plant Industria Ballkanike 2 sh.p.k were about 500,000 EUR (see Table 6.34). During the interview with the president of Industria Ballkanike, he highlighted several times his commitment to operate in premises that meet all the European Standards.

The major share of investments in operating facilities consist in the buildings in the two production plants with 500,000 EUR in Industria Ballkanike 1 sh.p.k and 250,000 EUR in Industria Ballakanike 2 sh.p.k. When the head office decided to open

Table 6.34 Investments in Industria Ballkanike sh.p.k (2015)

Category of investment	Amount in EUR
Investments in operating facilities	
Building	750,000
Ventilation	80,000
Electricity generators	230,000
Guest house/canteen	150,000
Security	80,000
Investments in machines	
Cutting	100,000
Sewing	400,000
Printing	100,000
Embroidery	70,000
Stamping	100,000
Pressing	30,000
Investments in information technology	
Hardware	90,000
Software	30,000
Total investments	2,210,000

Source Based on the information provided by Industria Ballkanie sh.p.k

the second production plant, based on local legislation it was able to privatize the previously state owned clothing manufacturing enterprise, which was operating for over 10 years under its potential capacity. The management of the enterprise decided to renovate production facilities and to locate part of its manufacturing activity in a second production plant.

The facilities of the enterprise are equipped with a ventilation system, which ensures an adequate temperature required during production of fabrics used in manufacturing of swimwear. Concurrently, there is a ventilation system, which absorbs the particles generated during sewing and cutting of goods. The total amount invested in ventilation system is 80,000 EUR.

The second major investment in operating facilities of Industria Ballkanike sh.p.k is made on electricity generators. Still in 2015, the region of Korca does not have uninterrupted power supply. Since its first days of operations in the host territory, the enterprise has had electricity generators, as frequent interruptions in the power supply would significantly harm the machinery used in manufacturing and the delivery of finished articles to end customer. Even during fieldwork in the premises of Industria Ballkanike, the electricity was interrupted twice. Investments made in electricity generators amount to 230,000 EUR. However, the maintenance of these generators is performed on a monthly basis to avoid any possible defects in them. The yearly amount spent on the maintenance of generators reaches to 20,000 EUR.

Premises of Industria Ballkanike sh.p.k have canteen services, which are used by employees during the day for lunch and coffee breaks. The enterprise does not have any guesthouse for representatives of customers or technicians that visit the two subsidiaries.

To continue, the expansion of Industria Ballkanike sh.p.k has occurred in parallel with the acquisition of modern machines. The majority of machines found in Industria Ballkanike sh.p.k are sewing machines: 250 in the first plant and 80 in the second plan. Sewing machines are acquired from brands like Juki and Gerber and are programmed to perform tasks like the use of multiple threads, different kinds of seams, ruffles, etc. The amount spent in sewing machines totals to 400,000 EUR.

In addition to sewing, Industria Ballkanike sh.p.k has six cutting machines that are used to cut the various types of fabrics automatically or manually depending on the category of the model that is under production. Cutting machines are programmed before a model enters into production. Investments in cutting machines amount to 100,000 EUR and it is expected to go up, as more customers require the services of Industria Ballkanike. In addition, Industria Ballkanike sh.p.k has sophisticated printing machines, which are used to print on large scale models received from customers while machines used in stamping are used to insert designs (stamps). The investment made on these machines amounts to 200,000 EUR in the two production plans. Another investment made in Industria Ballkanike sh.p.k is on embroidery machines in which the employer inserts the model through a flash drive to insert the logo of each customer. Finally, production in Industria Ballkanike sh.p.k is facilitated through the presence of additional machines used for stitching, pressing, buttoning, etc. required to meet the needs of its customers.

Industria Ballkanike sh.p.k has made also investments in information systems. The administration is equipped with computers supported with servers and software used to monitor the level of inventory of imported raw materials and exports of finished goods. In the same time, modern programs and specialized software for preparing the designs received from the customers are available in the subsidiary. Additional investments are made in computer hardware that are placed in several points in production facilities and are used to track all stages of production for each article manufactured in the subsidiary. These computers operate under a specific program, and are directly linked to the computer of the heads of various divisions and the production manager.

At the beginning of its activity, Industria Ballkanike offered only sewing services to its customers. Under the guidance of its founders, more processes and production lines started to operate in the premises of the two production plants. Industria Ballkanike is capable to meet all international standards (ISO) required by its customers (see Fig. 6.36).

During fieldwork in the two production plants the management, supervisors, and shop floor workers emphasized that production in Industria Ballkanike is guided by team work between the management and production at all levels. Production starts immediately after the president of Industria Ballkanike has agreed with customers on the terms and services to be provided. Industria Ballkanike receives raw materials selected by each customer. In parallel, the subsidiary gets electronically the

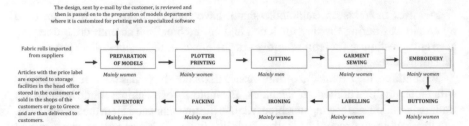

Fig. 6.36 Production flow (In the production flow embroidery and buttoning as showed with interrupted lines as they are subject only for specific articles in Industria Ballkanike sh.p.k) in Industria Ballkanike sh.p.k. *Source* Author representation based on information provided by Industria Ballkanike sh.p.k

design together with technical specifications for samples so that various processes like cutting, sewing, embroidery, control of defects, cleaning, ironing, etc. can start. After samples are prepared and sent to customers, they determine the quantity and the delivery date of the order. Among the customers, Decathlon has an office in Industria Ballkanike sh.p.k in which two technicians monitor the production of their output.

The overall output of Industria Ballkanike has increased by 61.54% within the last four years amounting to 210,000 items per year (see Table 6.35). The main category in the output of Industria Ballkanike is bikini with 65% in the total output of 2014. The smallest share in the output of Industria Ballkanike is on children line with only 3%.

The management of Industria Ballkanike is currently working towards two new expansion projects. The first consists in the construction of a new production line for Decathlon within the existing premises. For the new production line 120 new workers will be employed. The other project expected to be completed within 5 years includes the opening of another subsidiary in the region. The management is considering two options. The first one is the city of Tetova across the border with Macedonia and the other option is the south of Kosovo.

Table 6.35 Output of industria Ballkanike in pieces (2010–2015)

No.	Item	Output (pieces) 2010	Output (pieces) 2015	Change in output (%)
1.	Bikini	78,000	136,500	75.00
2.	Swimsuits	26,000	31,500	21.15
3.	Shorts	19,500	21,000	7.69
4.	Shirts	10,400	14,700	41.35
5.	Children line	2,600	6,300	142.31
6.	Total	130,000	210,000	61.54

Source Based on the information provided by Industria Ballkanike

After consolidating its presence in Albania with the establishment of two production plants Industria Ballkanike does not intent to source production to subcontractors. It expects to expand its existing modern facilities or to open another subsidiary
in the region. According to the management of Industria Ballkanike, the enterprise
aims to provide full package services to its customers. Industria Ballkanike is a subcontractor of Shqiperia Trikot sh.p.k when it needs to manufacture its swimwear.
Their cooperation will last until Shqiperia Trikot sh.p.k finalizes its own production
line for swimwear.

This section presents the level of employment in the two production plants of
Industria Ballkanike, the training programs for employees at all organizational layers,
and their compensation.

In 2015, Industria Ballkanike had a staff of 529 employees in the two production
plants out of which 493 are female and 36 are male (see Table 6.36). In peak seasons
especially between April and August the subsidiary hires between 35 and 40 seasonal
employees. Male employees including the president, the general administrator, the
responsible for inventory, and the responsible for packing manage the subsidiary.

Table 6.36 Employment in industria Ballkanike (2015)

Plant		Industria Ballkanike 1			Industria Ballkanike 2			Totals
No.	Category	Total	Female	Male	Total	Female	Male	Value
1.	Administration	27	10	17	10	5	5	37
2.	Sewing lines	218	217	1	70	70	0	288
3.	Preparation of Models	12	12	0	0	0	0	12
4.	Quality Control in Production	27	27	0	9	9	0	36
5.	Cleaning of sewing remains	17	17	0	8	8	0	25
6.	Packaging	26	23	3	10	10	0	36
7.	Inventory	6	6	0	0	0	0	6
8.	Quality control of raw materials	5	5	0	0	0	0	5
9.	Ironing	11	11	0	0	0	0	11
10.	Cutting	2	0	2	16	10	6	18
11.	Responsible	17	17	0	6	0	0	23
12.	In charge for defects	7	7	0	0	0	0	7
13.	Stamping	5	5	0	0	0	0	5
14.	Embroidery	0	0	0	5	3	2	5
15.	Assistants	11	11	0	4	4	0	1
16.	Totals	391	368	23	138	125	13	529

Source Based on the information provided by Industria Ballkanike sh.p.k

Female employees dominate the three production lines and the administration. The Greek administrators are responsible mainly for (i) suggesting new techniques to increase staff productivity, (ii) setting future targets for the subsidiary, and (iii) providing solutions to the problems occurring during production that cannot be solved by local administration. In the second production plant there is one male employee operating a sewing machine and two that run the embroidery machines. Overall, male dominate in management positions with 22 employees while females are 15.

Proper maintenance of production facilities is guaranteed through 15 employees including mechanics, drivers, guards, gardeners, and cleaners. Mechanics are key personnel in daily operations of Industria Ballkanike. Each of them has over 15 years of experience and is able to fix on the spot the majority of problems occurring during production.

The criteria for selection of employees in Industria Ballkanike depends on the degree of responsibility and the difficulty of assigned tasks. The president has an advanced university degree, speaks at least two foreign languages, and possesses deep knowledge on clothing manufacturing. The general director possesses also has a good understanding of the business environment in the country. The manager of production is selected by the administrators of Industria Ballkanike and is approved by the president. Employees responsible for divisions and units are selected from the production manager of Industria Ballkanike based on previous experiences in the clothing industry. They need to have at least a secondary degree, speak one foreign language, and possess basic knowledge of computer software. In total, there are 23 division heads in Industria Ballkanike. The staff turnover rate is less than 2%.

Evaluation of employees is critical for Industria Ballkanike as it is linked to the quality of finished articles. Evaluation criteria varies by position, responsibility, and duties required to perform. For low to medium level positions occupied mostly by shop floor employees evaluation criteria include the number of goods completed daily, team-work, precision in goods manufactured, and amount of overtime hours spent to generate the assigned daily quantity. Employees responsible for units and divisions are evaluated based on the ability to instruct and guide employees they supervise, the ability to meet assigned targets, reliability, and the overall performance of the unit. The evaluation form is prepared by the production manager and is approved by the president and the administrator of the subsidiary. Employees that on a regular basis receive evaluations above the average have the opportunity to be promoted. For example, after several years employees that start working in sewing lines can become responsible for the unit or the division.

Industria Ballkanike is committed to provide adequate training to its employees. Trainings are organized for all level of employees. In the subsidiary, there is a special line to train new employees for 90 days. In the same time, all employees are trained to use a software to record the processes, the number, and the types of article they are working with so that the supervisors can monitor at any time at what state a single good is. Employees are trained on various processes like sewing, cutting, buttoning. At the end of the training period, employees are evaluated by their supervisors and depending on their performance are assigned to respective departments.

Continuous training is offered to employees of Industria Ballkanke form technicians of Decathlon.

Employees in charge of production units have had previous experience in clothing manufacturing. In total 20 employees have been exposed to international training especially by those that have been organized by Decathlon and C&A. These employees have gained extensive knowledge and practice in clothing manufacturing. Employees responsible for financial management are trained for using accounting programs required to record financial transactions of the subsidiary and for determining the monthly salary of each employee depending on the number of items produced daily. Trainings in Industria Ballkanike sh.p.k have included safety and health in the workplace, maintenance of sewing machines, training against discrimination of color, religion and origin.

Working hours at Industria Ballkanike are from Monday to Saturday between 8:00 am and 16.30 p.m. including a 30-min lunch break. The lunch break varies from one unit to the other. Employees are required to use the personal identification card to enter or to leave the company in order to record their working hours. The number of weekly working hours is in full compliance with the national Labor Code, a maximum of 50 h per week. Overtime hours during weekdays are paid 25% more, while working hours during national holidays are paid 125%. National holidays can be replaced with another day after approval with the supervisor or can be added to the annual leave of employees.

The annual leave for each employee is 28 working days including Saturdays and Sundays. All employees take 14 days of the annual leave on August when production is interrupted. The rest of the annual leave can be taken based on the needs of each employee. All employees are required to follow the set of safety and emergency rules and they all have their special uniform prepared depending on the responsibilities and tasks of each employee.

Employees in production are paid depending on the quality and quantity of goods they manufacture daily (see Table 6.37).

Table 6.37 Compensation of employees in Industria Ballkanike sh.p.k (2015)

No.	Category of employee	Monthly wage range (EUR)
1.	Administration	500
2.	Head of department/division	1,000
3.	Blue collar employees	200–250

Source Based on the information provided by industria Ballkanike sh.p.k

Chapter 7
Case Study Analysis of Foreign Clothing Manufacturing Enterprises

Abstract This chapter presents a detailed analysis of the four clothing manufacturing subsidiaries with regard to their evolution in the host territory for the last twenty years. The analysis offers a dynamic perspective as it deals primarily with the evolution of these subsidiaries with regard to the knowledge transferred in the host territory and the evolution in the quality of the subsidiary with regard to operational upgrading and embeddedness in the host territory.

Keywords Case study analysis · Evolution of the subsidiary · Knowledge transfer · Quality of the subsidiary

7.1 Knowledge Transfer in the Host Territory

With reference to the framework presented in Chap. 5, this section looks into the knowledge the four clothing manufacturing subsidiaries have transferred into the host territory. In order to determine the knowledge transferred particular attention is given to: (i) the quality of the group to which the subsidiary belongs, (ii) the stock of knowledge, and (iii) the channels used by clothing manufacturing subsidiaries to transfer knowledge in Albania (see Table 7.1).

The operational activity of the group directly affects the activity of the subsidiary in the host territory and therefore the knowledge transferred. In two of the case studies the majority of high value added operations are undertaken in the head office of the group. This is true for Naber Moden and Valcuvia s.r.l. While in Cotonella S.p.A and Industria Ballkanike the majority of operations are undertaken in their respective subsidiaries located in a developing country like Albania. With regard to the nature of operations, the group is responsible for key operations like marketing of own brand, design, relations with customers and suppliers, and market research. Similarities among the case studies occur also in the strategy of the group on the operations of the subsidiary. This strategy is mostly oriented toward allocating production to subsidiaries affecting this way the level and the degree of knowledge transferred in the host territory. An exception occurs with Cotonella S p A, which has assigned the subsidiary with additional operations including brand marketing, market research, and keeping a certain degree of relations with customers

© Springer Nature Switzerland AG 2020

J. Kacani, *A Data-Centric Approach to Breaking the FDI Trap Through Integration in Global Value Chains*, Lecture Notes on Data Engineering and Communications Technologies 50, https://doi.org/10.1007/978-3-030-43189-1_7

Table 7.1 Knowledge transferred in the host territory from the four case studies

Quality of the group				
Case	Shqiperia Trikot sh.p.k (Cotonella S.p.A)	Naber Konfeksion sh.p.k (Naber Moden)	Valcuvia Alba sh.p.k (Valcuvia S.r.l)	Industria Ballkanike sh.p.k (Industria Ballkanike)
Types of knowledge				
Operations	The head office is responsible for key operations like brand marketing, relationship with suppliers of raw materials, customers and market research Fractions of these activities are performed in the subsidiary	The head office is responsible for key operations like brand marketing, relationship with suppliers of raw materials, customers and market research	The head office is responsible for key operations like brand marketing, relationship with suppliers of raw materials, customers and market research	The head office is responsible for key operations like brand marketing, relationship with suppliers of raw materials, customers and market research
Production modality	Full package with design with production occurring in the subsidiary and subcontractors. It has also its own brand	Full package with design with production occurring in the subsidiary. It has also its own brand	Full package with design with production occurring in the subsidiary	Cut, make, trim
Strategy towards the subsidiary	Cooperation with the subsidiary in undertaking key operations and in planning additional investments to expand production activity	The subsidiary depends on the head office for its activity and is in charge only for production	The subsidiary depends on the head office for its activity and is in charge only for production	The subsidiary depends on the head office for its activity and is in charge only for production
Technical	Compulsory training for all employees for three months Ongoing and customized training for each unit On-site support from Italian technicians and experts Cooperation with clothing industry associations and universities Training of employees working for subcontractors	Compulsory training for three months Long-term training for engineers in the head office On-site support from German experts and technicians Drafting and promotion of a vocational training program for a professional school in the city of Durres Training of employees working for subcontractors until the termination of cooperation	Compulsory training for six weeks On-site support from Italian experts and technicians that run the subsidiary Organization of workshops in the subsidiary covering all departments. Customized trainings for specific units	Compulsory training for three months for each employee On-site technical support from technicians placed by major customers. (e.g. Decathlon) in operating facilities of the subsidiary On-site support from Greek employees that run the subsidiary

(continued)

Table 7.1   (continued)

Quality of the group

Case	Shqiperia Trikot sh.p.k (Cotonella S.p.A)	Naber Konfeksion sh.p.k (Naber Moden)	Valcuvia Alba sh.p.k (Valcuvia S.r.l)	Industria Ballkanike sh.p.k (Industria Ballkanike)
Managerial	Executive trainings for the management (high and middle) level occur in the premises of the subsidiary and the head office Organization of workshops on how to handle major customers Trainings for the management and employees of subcontractors	Executive trainings for high-level management occur in the head office Engagement of consultants like Sequa and Weiss Organization of workshops on how to handle major customers Until 2014 trainings for subcontractors	Executive trainings occur in the head office for high-level management Trainings are mostly for Italian employees that run the subsidiary Few local employees are trained in the head office	Executive trainings for high-level management occur in the premises of the subsidiary and in the head office of key customers (e.g. Decathlon). Organization of workshops from customers in the premises of the subsidiary Few local employees are trained in the head office

Channels of knowledge transfer

Via employees	No presence of foreign employers since the opening of the subsidiary in Albania Local employees range from high-level management to assembly workers	German employees (owners of Naber Moden) are the administrators of the plants in Albania Local employees only lately have been assigned to high-level management positions All departments and units involved in assembly are occupied by local employees	Italian employees appointed by the head office are the current administrators and the highest ranked managers of the subsidiary Local employees only lately have been assigned to high-level management positions. All departments and units involved in assembly are occupied by local employees	Greek employees (one of the owners) are the administrators together with other Greek employees as high level managers of the subsidiary Local employees only lately have been assigned to high-level management All departments and units involved in assembly are occupied by local employees
Linkages with the host territory	No linkages with local suppliers as they do meet the requirements of customers As part of decentralization five local subcontractors were established with employees formerly working for Shqiperia Trikot sh.p.k	No linkages with local suppliers as they do meet the requirements of customers The enterprise cooperated with five local subcontractors until 2014. Since January 2015 it does not have any local subcontractors	No linkages with local suppliers as they do meet the requirements of customers No current local subcontractors. The enterprise is looking for subcontracting companies for stamping purposes of different fabrics and models	No linkages with local suppliers as they do meet the requirements of customers No local subcontractors as they do not meet the requirements of Decathlon

(continued)

Table 7.1   (continued)

Quality of the group				
Case	Shqiperia Trikot sh.p.k (Cotonella S.p.A)	Naber Konfeksion sh.p.k (Naber Moden)	Valcuvia Alba sh.p.k (Valcuvia S.r.l)	Industria Ballkanike sh.p.k (Industria Ballkanike)
Demonstration effects	Occurring through establishment of five subcontractors that have expanded the activity of their enterprises by working with other clients other than Cotonella Few employees that have left Shqiperia Trikot sh.p.k and have established their own clothing manufacturing enterprises	Subcontractors have been able to operate on their own and to work directly with customers including those of Naber Moden and do not to depend exclusively on the orders received from Naber Moden Through employees that decided to create their company after Naber Moden acquired Trumph Blousen	Limited demonstration effects due to the absence of local subcontractors and limitations in employment of local workers in high management positions	Limited demonstration effects due to the absence of local subcontractors and limitations in employment of local workers in high management levels Demonstration mostly through employees that decided to move to other clothing manufacturing enterprises

Source Shqiperia Trikot sh.p.k, Naber Konfeksion sh.p.k, Valcuvia Alba sh.p.k, and Industria Ballkanike sh.p.k

and suppliers. In all case studies, the services in the subsidiary are offered to international customers while in the case of Cotonella S.p.A and Naber Moden services are rendered also for production of own brand.

Suppliers of raw materials are key for the operational activity of the group and their subsidiaries. The four groups have established developmental linkages with suppliers of raw materials. These groups have a long-term interaction with suppliers occurring mostly through periodic meetings. In all cases, these meetings encourage mutual learning like on how to produce new fabrics demanded by a customer or for manufacturing a specific collection for their own brand as in the case of Naben Moden and Cotonella S.p.A.

Cotonella S.p.A, Naber Moden, and Valcuvia s.r.l cooperate with subcontractors outside the host territory where their subsidiaries are located. However, the level of output generated by these subcontractors is very low ranging from 1 to 1.5% and occurs during seasonal production peaks when subsidiaries in Albania are unable to accommodate excess production of final goods. The impact of these subcontractors in the operations of the group is not significant.

7.1.1   Stock of Knowledge of the Subsidiary

In the host territory, foreign owned enterprise can transfer two kinds of knowledge (i) technical and (ii) managerial knowledge. The four clothing manufacturing enterprises differ among each other on the kinds of knowledge transferred in the host territory affected by the national groups they belong.

Technical knowledge is the most common knowledge transferred in the host territory. All clothing manufacturing subsidiaries have introduced a compulsory training program for their employees especially for those providing assembly services. The compulsory training ranges from six weeks in Valcuvia Alba sh.p.k to three months in the three remaining subsidiaries. Additional technical knowledge is transferred in the host territory through on-site interaction of foreign technicians (Italian, German, Greek, and French) with local employees during their frequent visits in production facilities in Albania to organize various workshops and seminars with international experts. The seminars are customized depending on the activity of each department/unit aiming to expose employees to more advanced knowledge than the knowledge acquired during the compulsory training. In case of Industria Ballkanike sh.p.k technicians of Decathlon are placed long-term in the subsidiary guiding local employees on a daily basis.

Technical knowledge is also transferred among various departments in the subsidiary. This occurs through staff rotation from one department to another where local employees exchange their know-how acquired while working in other departments.

However, only two subsidiaries managed to transfer additional technical knowledge in Albania. Naber Konfeksion sh.p.k has drafted and launched a vocational training program to prepare students according to the needs of the clothing industry and to create a potential base of qualified labor force in the region of Durres. This program started in 2016 to train the first generation of students. In addition, Shqiperia Trikot sh.p.k cooperates with industry associations including the Chamber of Commerce, the Austrian Development Agency and universities, especially with the Polytechnic University of Tirana, in order to organize trainings not only for its employees but also for its five subcontractors.

With reference to managerial knowledge the level transferred in the host territory has been limited and only for few employees in middle and high level management running the clothing manufacturing subsidiaries. In the four case studies, local employees have been able to gain managerial knowledge from executive trainings organized in the premises of subsidiaries and participation in multiple workshops led by foreign experts specialized in the provision of consulting services in the clothing industry. Workshops organized by international experts are mostly oriented towards handling major customers, receiving orders from international suppliers, and achieving better process flow in the operations performed within the subsidiary. Among the four case studies, Naber Konfeksion sh.p.k is the subsidiary that has cooperated mostly with international consultants including Sequa and Weiss from which local employees have benefited not only additional managerial knowledge but also know-how on establishing internal organization structures to achieve a smoother functioning of the subsidiary.

Managerial knowledge acquired in the head office has been less frequent as trainings organized in the head office have been very few and mostly for managers that have been working for many years in the subsidiary or for employees that have worked in different departments.

However, the knowledge transferred remains mostly within production function without involving more value added functions like design, research and development as they still remain within the head office. Little managerial knowledge is transferred as few local employees are appointed in management positions especially in high management. The only exception occurs in Shqiperia Trikot sh.p.k in which managerial knowledge is acquired also through relationships with customers, suppliers, brand promotion, and market research. In addition, managers in Shqiperia Trikot sh.p.k have acquired knowledge on preparing business plans with particular focus on new investments to expand operations and production activity of the subsidiary.

7.1.2   Channels of Knowledge Transfer

This section presents an analysis of the channels through which clothing manufacturing subsidiaries transfer knowledge into the host territory.

The first channel of technology transfer is through hiring of local employees in high management position to act as administrators or general managers of the subsidiary, local employees in middle management responsible for various departments, in low management responsible for various divisions the departments, and in production as shop floor workers.

In Shqiperia Trikot sh.p.k only local employees are appointed in the management and production since the start of operational activity in Albania. Local employees are appointed as the administrator of the subsidiary as department heads and responsible for divisions. As production expanded, the number of local employees serving in high management positions within Shqiperia Trikot sh.p.k increased by training and promoting employees that started to work in assembly since the early years of operations in Albania. What has favored the presence of local employees in the management of Shqiperia Trikot sh.p.k is the confidence of the Italian president and owner of Cotonella S.p.A in the abilities of local employees and his strong belief that in a host territory an enterprise is better run by local employees.

A different scenario occurs in the remaining three clothing manufacturing subsidiaries. Naber Konfeksion sh.p.k is administered by German employees that is also the owner of the three production plants in Albania. In Naber Konfeksion sh.p.k only recently local employees have been appointed in the high management of the subsidiary serving as the administrator and the manager but only for production of goods. However, local employees serve in middle management as department heads and low management as responsible of divisions throughout the presence of Naber Konfeksion sh.p.k in the host territory. In Industria Ballkanike sh.p.k one of the owners serves as the general administrator of the subsidiary and is assisted by two local employees that are appointed as the administrator and the manager for production of goods. In Valcuvia Alba sh.p.k the two administrators are directly appointed from the head office. Up to now, only one local employee has made it into the high management serving as the local administrator of the subsidiary. With regard to middle and low managers in Industria Ballkanike sh.p.k and Valcuvia Alba sh.p.k only local employees occupy these positions.

A common feature in the four clothing manufacturing subsidiaries is that local employees in high management have worked for the enterprise for at least fifteen years, have participated in multiple trainings both in the head office and in the premises of the subsidiary, and are fluent in the native language of the owners of the subsidiary. Employees appointed in middle management are promoted based on the performance they have achieved during multiple years of service in the subsidiary after starting as employees in assembly functions. A key factor that has limited the number of local employees in highly ranked management positions is the control that owners of the group want to have in the operational activity of the subsidiary and the time they need to build confidence in local employees. On the other hand, in the four clothing manufacturing subsidiaries, only local employees are present in low value added activities including production. Production consists mostly in assembly work. As previously presented value added activities like research and development, marketing, design, etc., remain with the head office and in three case studies, no efforts are made from the owners of the group to transfer them in Albania. What discourages owners to transfer these activities is the lack of expertise of local employees in performing such processes by local employees in Albania. An exception is Shqiperia Trikot sh.p.k in which a fraction of these functions is taking place Albania.

However, even in Shqiperia Trikot sh.p.k only few local employees go beyond mere assembly work. They are involved in more advanced processes like testing of raw materials, quality control of finished articles, participate in head office strategy meetings on selection of suppliers, are involved in brand marketing in the host territory, and perform minor research and development tasks on market analysis for promotion of the Cotonella brand in the region. Local employees in Shqiperia Trikot sh.p.k have been able to do so as the head office decided to decentralize production by transferring assembly activities to five subcontractors switching its priorities in training of local employees on more complex functions currently undertaken within the subsidiary.

The second channel through which knowledge is transferred in the host economy is through linkages established with local suppliers of intermediary goods and subcontractors. To start with, linkages established with the host territory, are influenced by the quality of the group they belong. With regard to Cotonella S.p.A and Naber Moden they are partially dependent on customers as their output is divided between their own brand and orders received from other customers. As such, they are independent from customers on the decisions they make on suppliers of raw materials and articles manufactured for Cotonella and Naber Collection. When it comes to orders from other customers, Cotonella S.p.A and Naber Moden are constrained to meet their requirements regarding raw materials and design.

On the other hand, Valcuvia s.r.l and Industria Ballkanike act only as subcontractors for international customers as they do not have their own brand. They are constrained to follow the requirements of the customers. The four clothing manufacturing subsidiaries have failed to establish any linkages with local suppliers. A unified statement comes from all clothing manufacturing subsidiaries that in the

Albanian market there are no local suppliers that can be contracted for provision of raw materials used as intermediate inputs in clothing manufacturing. Local suppliers are absent even for simple raw materials like plastic bags and card boxes because they do not meet the requirements of imposed by customers in terms of quality and price or the brand managed by the group.

Differently from linkages with suppliers of raw materials, linkages with subcontractors occur for specific case studies both at a group and at subsidiary level. With regard to linkages with local subcontractors, the four case studies display a mix behavior. Shqiperia Trikot sh.p.k has managed to create strong linkages as its five subcontractors are managed by employees formerly working and trained in Shqiperia Trikot sh.p.k. The linkages of Shqiperia Trikot sh.p.k with its subcontractors are mostly dependent in nature. Four of its subcontractors rely entirely on the orders they receive from Shqiperia Trikot sh.p.k while only one is semi-dependent with only 50% of its orders coming from Shqiperia Trikot sh.p.k. Despite of the technical support and training offered by Shqiperia Trikot sh.p.k the operational activity of the subcontractors remains limited only to assembly functions without including additional ones like quality control or even cutting. Subcontracting of assembly services allows Shqiperia Trikot sh.p.k to expand its operational activity on more advanced functions like marketing and research and development.

To continue, the nature of the linkages of Naber Konfeksion sh.p.k with its subcontractors has fluctuated since the start of production activity from fully dependent to completely independent. In the early 1990s Naber Moden was the only customer of local subcontractors providing mostly assembly services. When Naber Moden decided to open its own production plants linkages with local subcontractors turned into semi-dependent as subcontractors started to produce also for other customers. Naber Konfeksion sh.p.k turned from the only customer of its local subcontractors into one of the customers, as with the experience they gained from manufacturing for Naber Konfeksion sh.p.k they were able to expand production activity. In early 2015, when the third production plant of Naber Konfeksion sh.p.k opened in Albania the owners decided to terminate any linkages with local subcontractors that operate independently. During the years of cooperation with local subcontractors, Naber Konfeksion sh.p.k provided technical support, however limited only to assembly services. This technical support proved beneficial for local subcontractors to become self-sufficient in provision of assembly services to international customers.

The remaining two subsidiaries Valcuvia Alba sh.p.k and Industria Ballkanike sh.p.k do not have any subcontractors in Albania, as they have not found any subcontractor to meet the requirements set by their customers. Only in 2014, Valcuvia Alba sh.p.k started to look for a subcontractor that can offer stamping services on a variety of fabrics used in manufacturing of cloths. Despite of these efforts, the administrators of Valcuvia Alba sh.p.k had not managed to find an appropriate local firm to rely on for the required stamping services. Another reason for the absence of local subcontractors is because these two subsidiaries are not in the capacity to provide full training services due to the absence of their own brand as the group it belonged operates at as a subcontractors to international brands.

The third channel of knowledge transfer occurs through demonstration effects. Demonstration effects in the four clothing manufacturing subsidiaries in the Albanian economy have been very limited and mostly occurring in Shqiperia Trikot sh.p.k and Naber Konfeksion sh.p.k. In Shqiperia Trikot sh.p.k the majority of demonstration effects occurred when it decided to reorganize production that lead to the establishment of five new local clothing manufacturing enterprises administered by former local employees of Shqiperia Trikot sh.p.k that provide assembly services. Also, demonstration effects have occurred in the establishment of local clothing manufacturing enterprises from former employees of Shqiperi Trikot sh.p.k that do not serve as subcontractors.

A similar situation occurs in Naber Konfeksion sh.p.k in which demonstration effects happened through five local manufacturing enterprises that previously served as subcontractors of Naber Moden. Today, these local clothing manufacturing enterprises are fully capable to operate and serve a number of customers, including some that they share with Naber Konfeksion sh.p.k. In addition, when Naber Moden decided to acquire Trumph Blousen some employees decided not to join the new owner but to establish new local clothing manufacturing enterprises. The other two subsidiaries, Valcuvia Alba sh.p.k and Industria Ballkanike sh.p.k have the lowest degree of demonstration effects as they did not have any subcontractors and are characterized by a very low degree of labor turnover. Therefore, no former employee has been able to open a local clothing manufacturing enterprise.

7.2 Subsidiary Upgrading in the Host Territory

Upgrading in the four clothing manufacturing enterprises is analyzed at the subsidiary level. As introduced in the framework of Chap. 5 it will focus on the evolution of activities realized overtime and the integration within the host territory. In addition, this section focuses on the factors that have influenced the different upgrading paths of subsidiaries in the host territory.

7.2.1 Subsidiary Upgrading: Complexity of Activities

To start with, three kinds of upgrading are analyzed based on the activities of the subsidiary and the integration within the host territory (see Tables 7.2. 7.3, 7.4 and 7.5).

Since early 1990s when the four clothing manufacturing subsidiaries started their activity in Albania they have achieved a certain degree of process upgrading mostly dominated by adding new processes in order to complete production. In the 1990s the processes performed in the subsidiaries were limited to cut, make, and trim. Gradually the four clothing manufacturing subsidiaries started to add more processes in

Table 7.2 Summary of evolution of Shqiperia Trikot sh.p.k

Subsidiary	1990		2015	
Case study	Complexity of activities	Integration within the host territory	Complexity of activities	Integration within the host territory
Shqiperia Trikot sh.p.k (Cotonella S.p.A)	Engaged only in assembly (CMT) Production of standardized articles like vest, slips, and bra for men and women	Only local employees in administration, high, middle, low management, and production No local suppliers as they do not meet the requirements for the brand and for customers No local subcontractors	Full package services with own brand Control of raw materials Deliver of finished articles to end customer Involvement in R&D projects especially in the Balkan region Marketing and brand promotion through opening of Cotonella brand shops in Albania and Kosovo Involvement in the selection of suppliers Increase in the range of standardized products including vest, slips, bra, thermal underwear, pajamas, night gowns for men, women, children Production of differentiated products of luxury line of swimwear Working with the head office in realizing the regional center for manufacturing of fabric and dyeing	Only local employees in administration, high, middle, low management, and production No local suppliers as they do not meet the requirements for the brand and for customers Five local subcontractors that offer assembly services

Source Shqiperia Trikot sh.p.k

Table 7.3   Summary of evolution of Naber Konfeksion sh.p.k

Subsidiary	1990		2015	
Case study	Complexity of activities	Integration within the host territory	Complexity of activities	Integration within the host territory
Naber Konfeksion sh.p.k Naber Moden	Production through subcontracting a local clothing manufacturing enterprise for assembly services (CMT) Production of standardized articles like shirts, dresses, skirts, etc.	Foreign employees in the administration and high management Local employees in middle and low management Local employees in production No plants in the host territory Plants and subcontracting in other territories No local suppliers as they do not meet the requirements for the brand and for customers Investments in acquiring a former German company and opening of the first plant in the host territory Local subcontractors offer assembly services	Full package services with brand Increase in the range of standardized. products manufactured including parkas, coats, trousers, jeans, blouses, etc. for men, women, and children Production of differentiated products like traditional dresses and expensive outfit for international brands like Versace Additional processes within the production function like buttoning, embroidery, inventory, etc. Key functions like marketing, research and development, selection of suppliers, etc. remain in the head office Dependency from the head office in the operational activity of the subsidiary Working with the head office staff training and establishment of a vocational training program	Foreign employees in the administration of the subsidiary Local employees in high, middle, low management and production Three plants in the host territory No local subcontractors after opening of the third production plant No local suppliers as they do not meet the requirements for the brand and for customers Considerable investments in operating facilities, machines, and information technology

Source Naber Konfeskin sh.p.k

their operational activity including quality control of raw materials, labeling, buttoning, ironing, etc. The only clothing manufacturing subsidiary that is able to realize processes that fall outside production is Shqiperia Trikot sh.p.k. The type of articles clothing manufacturing subsidiaries produce also affects process upgrading. Naber Konfeksion sh.p.k and Industria Ballkanike sh.p.k are able to offer more specialized

Table 7.4 Summary of evolution of Valcuvia Alba sh.p.k

Subsidiary	1990		2015	
Case study	Complexity of activities	Integration within the host territory	Complexity of activities	Integration within the host territory
Valcuvia Alba sh.p.k (Valcuvia s.r.l)	Only assembly services. Production of standardized articles slips and shirts for men and women	Foreign employees in the administration and high management Local employees in middle and low management Local employees in production No local suppliers as they do not meet the requirements for the brand and for customers No production plants Renting of old buildings	Full package services Increase in the range of standardized products manufactured including nightgowns, pajamas, thermals, etc. for men, women, and children Production of differentiated products like sport outfit for international cycling competition Additional processes within the production function like stitching, molding, inventory, etc. Key functions like marketing, research and development, selection of suppliers, etc. remain in the head office Dependency from the head office in the operational activity of the subsidiary. Working with the head office on expansion of production facilities	Foreign employees in the administration of the subsidiary Local employees in high, middle, low management and production One plant in the host territory No local subcontractors No plants in other host territories Subcontracting in other host territories on seasonal basis No local suppliers as they do not meet the requirements for the brand and for customers Considerable investments in operating facilities, machines, and information technology

Source Valcuvia Alba sh.p.k

Table 7.5 Summary of evolution of Industria Ballkanike sh.p.k

Subsidiary	1990		2015	
Case study	Complexity of activities	Integration within the host territory	Complexity of activities	Integration within the host territory
Industria Ballkanike sh.p.k (Industria Ballkanike)	Only assembly services Production of standardized articles, sportive shirts and trousers for men and women	Foreign employees in the administration and high management Local employees in middle and low management Local employees in production No local suppliers as they do not meet the requirements for the brand and for customers. No production plan Renting of old buildings	Cut, make, trim Increase in the range of standardized products manufactured including shirts, trousers, swimwear, etc. for men, women, and children Additional processes within the production function like embroidery, inventory, etc. Key functions like marketing, research and development, selection of suppliers, etc. remain in the head office Dependency from the head office in the operational activity of the subsidiary Working with the head office on expansion of production facilities with another plant in a neighboring country	Foreign employees in the administration of the subsidiary Local employees in high, middle, low management and production One plant in the host territory No local subcontractors No plants in other host territories Subcontracting in other host territories on seasonal basis No local suppliers as they do not meet the requirements for the brand and for customers Considerable investments in operating facilities, machines, and information technology

Source Industria Ballkanike sh.p.k

processes within production like embroidery, as they are required to manufacture traditional dresses or symbols in sport outfits. The other two subsidiaries Shqiperia Trikot sh.p.k and Valcuvia Alba sh.p.k, oriented in manufacturing of underwear, realize more specialized processes within production that refer to ultrasound procedures on various underwear required by customers and stamping of fabric used in manufacturing of cycling outfit.

Product upgrading has been limited in the four clothing manufacturing subsidiaries as the majority of articles manufactured in Albania are categorized as standardized goods mostly for wide spread consumption at reasonable prices. Even though all manufacturers have more than twenty years of experience in the clothing industry in Albania, differentiated goods manufactured in the host territory in 2015 are a few and include: (i) traditional German dresses, (ii) luxury goods like fur articles and branded outfits (Versace, Calvin Klein, etc.) and (iii) sports outfits produced for international cycling competitions.

However, all four clothing manufacturing subsidiaries within the range of standardized articles have been able to produce more sophisticated items. Naber Konfeksion sh.p.k started with manufacturing of cotton shirts and today is able to produce a variety of blouses made of delicate fabric like silk. Shqiperia Trikot sh.p.k and Valcuvia Alba sh.p.k have been able to produce thermal underwear, pyjamas on any fabric, and swimwear. Industria Ballkanike sh.p.k started to manufacture only shirts for Decathlon and today is able to produce running uniforms and swimwear.

Functional upgrading in the clothing manufacturing subsidiaries has been very limited as the head office still remains responsible for more advanced operations like design, marketing, selection of suppliers, research and development, etc. Only Shqiperia Trikot sh.p.k and Naber Konfeksion sh.p.k have been able to achieve a minor functional upgrading since the start of their activity in the early 1990s. The minor functional upgrading derives from the ability of these two subsidiaries to expand their operational activity beyond production. To start with, Shqiperia Trikot sh.p.k has been responsible for opening of "Cotonella" brand shops in Albania and Kosovo and it has been involved in research and development projects in the Western Balkan countries to conduct a market analysis with the ultimate goal to expand the market share of "Cotonella" brand in the Balkans.

In addition, Shqiperia Trikot sh.p.k has been able to create its own production line of swimwear sold in the United States. Naber Konfeksion sh.p.k has achieved minor functional upgrading mostly from the involvement of the subsidiary in undertaking research on identifying the gaps in the educational system. Naber Konfeksion sh.p.k has proposed a new training program for a vocational training school in the region of Durres in order to prepare the students in line with the needs of the clothing industry. Within Naber Konfeksion sh.p.k a slight involvement in functional upgrading has occurred through engagement of textile engineers in the design of various blouses for Naber Moden Collection that sells successfully in the shops throughout Germany.

On the other hand, the two remaining subsidiaries Valcuvia Alba sh.p.k and Industria Ballkanike sh.p.k have not been able to obtain functional upgrading as their operating activity still remains within production and does not involve any marketing, design, or selection of suppliers. Industria Ballkanike sh.p.k and Valcuvia sh.p.k are far from functional upgrading, as its activity is dominated by assembly services.

7.2.2 Subsidiary Upgrading: Integration Within the Host Territory

With regard to integration within the host territory process upgrading is achieved by delegating mostly assembly processes to local subcontractors. Local subcontractors that perform such processes are still present only in Shqiperia Trikot sh.p.k. In Naber Konfeksion sh.p.k integration in the host territory decreased since it stopped to place orders to local subcontractors. Integration of Valcuvia Alba sh.p.k and Industria Ballkanike sh.p.k is limited to their operational activity, as they did not have any subcontractors since the start of manufacturing activity in Albania.

Subsidiary upgrading in the sense of integration in the host territory for all case studies is not achieved as none of these subsidiaries has local suppliers that can provide raw materials to manufacture products they assemble. Local suppliers are not present as none of them is able to provide raw materials according to the standards required by international customers of the clothing manufacturing subsidiaries.

Table 7.3 presents a summary of the evolution occurring in the four subsidiaries after installation of production in the host territory. All subsidiaries have experienced evolution. The evolution reflects the characteristics of the group they are part of. Key functions like design, selection of suppliers, marketing, etc. are performed in the head office.

For each case study, a table is prepared to synthesize the evolution experienced after settling of production activity in Albania. In order to analyze each case a comparison is made on the operational activity of the subsidiary between early 1990s up to 2015.

7.3 Factors Affecting Subsidiary Upgrading in the Host Territory

This section presents the factors that affect upgrading at the subsidiary. These factors explain the different upgrading paths in the four clothing manufacturing subsidiaries.

7.3.1 Head Office Assignment

As seen in the literature and the framework, head office assignment is a key factor that affects subsidiary upgrading. In the four case studies, the degree of independence between the clothing manufacturing subsidiaries in Albania and respective head offices varies from case to case. The highest degree of independence is observed between Shqiperia Trikot sh.p.k and its head office Cotonella S p A Among the four case studies, Shqiperia Trikot sh.p.k is the only subsidiary to which the head office has assigned responsibilities that go beyond production. Their relationship is oriented towards a high degree of trust in local management of the subsidiary. The head office has appointed Shqiperia Trikot sh.p.k in charge of 98% of total output and has engaged the subsidiary: (i) to participate in the meetings organized in the premises of the head office aiming at selection of suppliers of raw materials including fabrics with a high content of cotton, (ii) to undertake regional research and development activities on market trends and preferences and (iii) to promote the "Cotonella" brand in the Balkan Peninsula. The freedom given to local employees appointed in high level management in reorganizing production in Shqiperia Trikot sh.p.k has resulted not only in the establishment of five subcontractors but has also proved beneficial in the identification of new business expansion opportunities like the upcoming investment to make functional a regional operational center for which a 12 million EUR investment is underway. It was the local management of Shqiperia Trikot sh.p.k that identified the need to open a regional center for production of fabric and dyeing facilities to serve the clothing manufacturing enterprises located in South East Europe. Together with the head office, the management of Shqiperia Trikot sh.p.k participated also in the preparation of the business plan of the regional center.

With regard to the three remaining clothing manufacturing subsidiaries, the degree of independence from the head office diminishes gradually. In Naber Konfeksion sh.p.k and its respective head office Naber Moden the degree of independence is lower than in the case of Shqiperia Trikot sh.p.k. In Naber Konfeksion sh.p.k local employees appointed in management positions are involved in the decision making process only by informing the owners during frequent visits in production facilities in Albania on the operational activity of the three production plants. Despite of this slight involvement, the head office has not taken any steps to delegate in Albania activities including design, marketing, purchase of raw materials, etc.

Finally, Valcuvia Alba sh.p.k and Industria Ballkanike sh.p.k are fully dependent for their operational activities from the head office. The respective head offices take all the decisions regarding production and operations on respective subsidiaries. The management in Valcuvia Alba sh.p.k and Industria Ballkanike sh.p.k whether foreign or local strictly follows all instructions and guidelines provided by the head office. Due to this dependency, these two subsidiaries within the last 15 years have not been able to perform any activity that goes beyond that of production function.

7.3.2   Local Environment

The four clothing manufacturing subsidiaries operate in a small country like Albania and are exposed to the same traits of local environment that affect their operational activity and upgrading. During fieldwork with the management in the four clothing manufacturing subsidiaries, the location[1] within Albania (different regions) did not

---

[1]Due to the small size of the host territory the region in which the subsidiary is located has an insignificant impact in the upgrading of subsidiaries.

influence their upgrading. Among local environment factors identified from field-work that hinders upgrading in the four clothing manufacturing subsidiaries are the unavailability of uninterrupted power supply and the lengthy custom procedures.

From on-site interviews, the most cited obstacle for production was the inability of Albanian authorities to provide uninterrupted power supply in the regions where these subsidiaries operate in particular and the country in general. Frequent interruptions in power supply is costly. In order to ensure uninterrupted power supply considerable investments are made not only in purchasing power supply generators but also in running and maintaining the equipment. The budget allocated yearly for ensuring uninterrupted power supply is an additional cost, which shrinks the funds available for additional investments and in the same time diminishes the confidence to obtain efficient production within the local environment. Moreover, frequent interruptions in power supply may damage sewing, cutting, buttoning, etc. machines and in most cases, machines need to be re-programmed. These difficulties encountered during production restrain the head office to bring more advanced machines in the host territory as they are afraid that these machines will damage quickly. Consequently, the head office is reluctant to bring additional operations in the country and to expose local employees to more advanced training. The other impeding factor identified during the fieldwork are the lengthy custom procedures in importing raw materials that are required to produce finished articles. The existing lengthy customs procedures increase the time required to deliver final goods. They pose a threat to the favorable geographical position, a major competitive advantage of Albania, on ensuring quick delivery times to end customers or the head office. In addition, overcoming lengthy custom procedures requires the involvement of the management of the subsidiary limiting their time to engage in additional activities or even to undertake in Albania more complex activities.

Another impeding factor consists in the frequent changes of the fiscal package that starting from 2010 are introduced every year in the Albanian economy. These changes bring uncertainty to the owners/administrators of the subsidiaries and to the head office on the decisions they need to make on future investments, expansion of production capacities, and the increase of more advanced technology in Albania. Frequent changes in the fiscal package mostly on taxes and subsidies are a threat to long-term investments, which require considerable amounts and consist in more advanced knowledge for the host territory.

The absence of economic zones is the final impeding factor identified during fieldwork in the four clothing manufacturing subsidiaries. The four cases identified the benefits they might have if production occurred in an economic zone where they could cooperate with other enterprises that could complement their production activity. These may include availability of local suppliers that meet the requirements of customers or other enterprises that may offer services for fabric dyeing that can facilitate the operational activity.

7.3.3   Investments in the Host Territory

This section introduces investments undertaken in the host territory. It focuses on the investment that increases the presence of the clothing manufacturing subsidiaries in the host territory.

The four clothing manufacturing subsidiaries have made considerable investments in the host territory (see Fig. 7.1). The majority of investment is in the form of fixed assets that consist mostly in production facilities including purchase of land and construction of buildings that are customized to fit the organizational arrangements and steps followed in production. The second type of investment made refers to machines used to realize production in Albania. All four clothing manufacturing enterprises have purchased modern sewing, cutting, embroidery, etc. machines that are present in production plants. For this type of investment, the clothing manufacturing enterprises have a risk to incur a loss due to the yearly depreciation of machines, fluctuations in the price of used machines, and transportation costs, as they cannot be easily moved from one territory to the next. The third category refers to additional investments that include information technology (computers, inventory system, etc.) within subsidiaries, vehicles used for transportation of employees, and power generators in order to ensure uninterrupted electricity during production.

The kinds of investments made are the same in all clothing manufacturing subsidiaries however they differ in the amount invested. Shqiperia Trikot sh.p.k is the subsidiary with the highest level of investment followed by Valcuvia Alba sh.p.k, Naber Konfeksion sh.p.k, and Industria Ballkanike sh.p.k. The higher the amount of investments the greater the presence to the host territory. This is particularly true for immovable investment like purchasing of land and operating facilities.

Investments in the four clothing manufacturing subsidiaries are continuous for two main reasons. Firstly, investments are made because of expansion in the production activity coming mostly from the increase in the orders received from customers. Secondly, investments are continuous as the head offices decided to close production

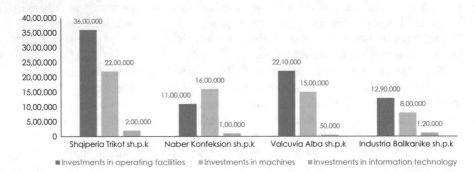

Fig. 7.1 Investments in four clothing manufacturing subsidiaries in Albania up to 2015 (EUR). *Source* Shqiperia Trikot sh.p.k, Naber Konfeksion sh.p.k, Valcuvia Alba sh.p.k, and Industria Ballkanike sh.p.k

plants in other countries and transfer production to Albania. The owners did not encounter many difficulties to close production plants in other host territories, as the level of investment made was not significant. They operated under rented facilities and subcontracted most of production activities to local enterprises. As previously shown major investments are made mainly in operating facilities and machines.

The fieldwork served also to identify upcoming projects planned by the four clothing manufacturing subsidiaries in Albania. Even though projects are different for each subsidiary, are categorized into: (i) expansions of production facilities, (ii) manufacturing of new articles, (iii) targeting of new customers, (iv) brand promotion, and (v) creation of a qualified labor force.

- Expansion of production facilities will occur in Shqiperia Trikot sh.p.k that intends to establish a regional production center with an investment expected to amount up to 12 million EUR. The center will produce various fabrics and will serve as large-scale factory for dye of various fabrics. In Valcuvia Alba sh.p.k and Industria Ballkanike sh.p.k new production facilities will be obtained by expanding existing facilities.
- Production of new articles is expected to occur in Valcuvia Alba sh.p.k as it will manufacture sports outfits mostly for international cycling competitions and will introduce a limited edition of female intimate apparel for the luxury German brand Schiesser. In addition, Shqiperia Trikot sh.p.k intents to expand its production line of swimwear.
- Targeting of new customers is a common objective in the four clothing manufacturing subsidiaries. Shqiperia Trikot sh.p.k aims to produce for Victoria Secret; Industria Ballkanike sh.p.k for Nike; Valcuvia Alba sh.p.k for Chantelle; and Naber Konfeksion sh.p.k for Calvin Klein Jeans.
- Brand promotion occurs only in Shqiperia Trikot sh.p.k as it is appointed by the head office to increase the number of shops selling the "Cotonella" brand products in Albania and in the Balkan region.
- Efforts to create a qualified labor force are made by Naber Konfeksion sh.p.k that has launched a vocational training program with professional schools in the region of Durres and by Shqiperia Trikot sh.p.k that through cooperation with universities organizes different training programs that include participants from numerous clothing manufacturing enterprises that operate in the region of Shkodra.

7.3.4   Plants in Other Host Territories

Possession of substitute plants or subcontracting in another country are among the factors that facilitate the transfer of production to another location.

The head offices of the four subsidiaries: Cotonella S.p.A, Valcuvia Alba s.r.l, Naber Moden, and Industria Ballkanike do not possess any substitute plants in other host territories. Absence of substitute production plants, has constrained the head offices to make more investments in Albania in order to expand their production facilities and to accommodate an increase in the level of output.

Absence of substitute plans has left room for the three clothing manufacturing subsidiaries to subcontract a fraction of their output in other countries. Cotonella S.p.A subcontracts part of production (pajamas and economic bras lines) in China and India as it can save on costs, as raw materials are ready available in these countries. Despite of subcontracting, the ultimate objective for Cotonella S.p.A is to transfer production from Asian countries to Albania. After the acquisition of "Linea Sprint", an international swimwear brand, Shqiperia Trikot sh.p.k started subcontracting also in Tunisia a fraction of the output due to limited capacities in the host territory.

Valcuvia s.r.l subcontracts a fraction of production in Tunisia due to limited production capacity it has in Albania. However, with expansion in production facilities that are underway Valcuvia s.r.l aims to transfer production from Tunisia into Albania (Table 7.6).

The enterprise will continue to subcontract in Egypt especially production of articles that require delicate fabric like lace and silk due to the availability of raw materials in the country. Industria Ballkanike sh.p.k is the only production facility of the enterprise. With regard to subcontractors abroad, the enterprise finds it more advantageous to produce in Albania where expansion of production facilities is underway.

Naber Moden subcontracts a fraction of its production in FYR Macedonia, as production in the plants in Albania could not accommodate all the orders received from customers. However, with the opening of the third production plant in Albania subcontracting in FYR Macedonia will soon terminate.

7.4   Conclusions

With reference to knowledge transferred in the host territory and the evolution in the quality, the four clothing manufacturing subsidiaries can be classified as presented in Fig. 7.2. The degree of integration within the host territory refers to the linkages subsidiaries have created with the host territory (suppliers and subcontractors), creation of a qualified labor force, and investments made. On the other hand, the complexity of activities refers to the kinds of upgrading achieved by each clothing manufacturing subsidiary.

- Shqiperia Trikot sh.p.k can be categorized as a complex embedded subsidiary as it is well integrated in the host territory coming mainly from linkages with local subcontractors, its efforts to create a qualified labor force through cooperation with various institutions (chambers of commerce, international organizations, and universities), and investments made in operational facilities and equipment. The degree of complexity of activities undertaken by Shqiperia Trikot sh.p.k is higher compared to the remaining subsidiaries as it has been able to obtain the three kinds of upgrading by expanding its operational activity beyond assembly services. The

Table 7.6 Factors affecting the quality of clothing manufacturing subsidiaries in Albania

Factors of upgrading				
Case	Shqiperia Trikot sh.p.k (Cotonella S.p.A)	Naber Konfeksion sh.p.k (Naber Moden)	Valcuvia Alba sh.p.k (Valcuvia S.r.l)	Industria Ballkanike sh.p.k (Industria Ballkanike)
Head office assignment				
Relationship with the head office	Close cooperation with the head office in the management of the subsidiary	Full dependency from the head office in running the subsidiary	Full dependency from the head office in running the subsidiary	Full dependency from the head office in running the subsidiary
Local environment				
Hindering factors	Unavailability of uninterrupted power supply Lengthy custom procedures Frequent changes in the fiscal regime Absence of economic zones	Unavailability of uninterrupted power supply Lengthy custom procedures Frequent changes in the fiscal regime Absence of economic zones	Unavailability of uninterrupted power supply Lengthy custom procedures Frequent changes in the fiscal regime Absence of economic zones	Unavailability of uninterrupted power supply Lengthy custom procedures Frequent changes in the fiscal regime Absence of economic zones
Subsidiary choice				
Level of subsidiary choice	The administrator and the management have some degree of freedom on taking decisions regarding operations of the subsidiary	The administrator and the management have no freedom to take decisions on the operations of the subsidiary	The administrator and the management have no freedom to take decisions on the operations of the subsidiary	The administrator and the management have no freedom to take decisions on the operations of the subsidiary
Other plants in other countries				
Possession of substitute plants in other countries	No possession of substitute plants in other countries	No possession of substitute plants in other countries	No possession of substitute plants in other countries	No possession of substitute plants in other countries
Subcontracting in other countries	Subcontracting in China, India, Tunisia	Subcontractors in FYR Macedonia	Subcontracting in Egypt and Tunisia	No subcontracting in other countries

(continued)

Table 7.6 (continued)

Factors of upgrading

Area	Shqiperia Trikot sh.p.k	Naber Moden sh.p.k	Valcuvia Alba sh.p.k	Industria Ballkanike sh.p.k
Investments in the host territory				
Investments	Considerable investments in owing operating facilities buildings and land Investments in technology (machineries + ICT) Investment in human resources of running the subsidiary from middle management and above Investments in distribution network Support subcontractors	Investments in owing operating facilities buildings and land Investments in technology (machineries + ICT) Investments in human resources especially middle management and above Investments in the local educational systems Investments for subcontractors	Investments in owing operating facilities buildings and land Investments in technology (machineries + ICT) Investments in human resources especially middle management and above Investments in electricity generators in which the factory is placed	Investments in owing operating facilities land and buildings Investments in technology (machineries + ICT) Investment in human resources especially middle management and above
Chronology in the investment	Repeated investments in operating facilities, machines, and information technology Investments are expected to increase in the future	Considerable investments in operating facilities, machines, and information technology Investments are expected to increase in the future	Repeated investments in operating facilities, machines, and information technology Investments are expected to increase in the future	Repeated investments in operating facilities, machines, and information technology Investments are expected to increase in the future

(continued)

Table 7.6 (continued)

Factors of upgrading				
Area	Shqiperia Trikot sh.p.k	Naber Moden sh.p.k	Valcuvia Alba sh.p.k	Industria Ballkanike sh.p.k
Upcoming projects	Establishment of a center to manufacture and dye fabric (12 mln/euro) Transfer in Albania of the production occurring in China, India, Tunisia in Albania Increase the number of Cotonella stores in Balkan region Production of swim suits Become a customer of Victoria Secret	Manufacturing of jeans through targeting of main brands of the jeans industry Preparation of the the first generation of students coming from vocational training according to the industry needs	Expansion of production site by starting to produce sport outfits especially in cycling Provision of stamping services to all customers. Establishment of a partnership with Schiesser in order to produce customized bras for seven models	Expand of production site Opening of another plant either in Albania or a neighboring country like Kosovo or FYR Macedonia

Source Shqiperia Trikot sh.p.k, Naber Konfeksion sh.p.k, Valcuvia Alba sh.p.k, and Industria Ballkanike sh.p.k in Albania

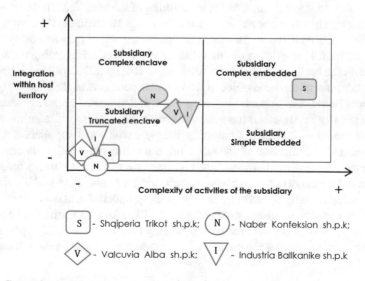

Fig. 7.2 Categorization of clothing manufacturing subsidiaries in Albania

responsibilities of Shqiperia Trikot sh.p.k go beyond production and include also marketing and selection of suppliers. With regard to the relationship with the head office, Shqiperia Trikot sh.p.k closely works with the head office in meeting the objectives of the group while it retains a certain level of independency in taking its own decisions in the daily management of the subsidiary and in proposing new organization structures or expansion projects. Shqiperia Trikot sh.p.k has also achieved a good degree of integration in the host territory through the presence of its subcontractors. Since the start of operations in the host territory, Shqiperia Trikot sh.p.k has not established any linkages with local suppliers.

- Naber Konfeksion sh.p.k is categorized as a complex enclave subsidiary. Its integration in the host territory results mainly from the qualified labor force it has created through training programs, establishment of clothing manufacturing enterprises with local ownership that previously were subcontractors of Naber Konfeksion sh.p.k, and investments made in making operational three production plants in Albania. On the level of complexity of activities realized in the host territory Naber Konfeksion sh.p.k has made little advancements to upgrade its operational activity beyond processes and activities involved in production. Referring to the relationship with the head office, Naber Konfeksion sh.p.k depends on the head office in undertaking most of the activities in the subsidiary. A good cooperation with the head office is observed in provision of consultancy services for training and reorganizing production processes of the subsidiary. However, the operational activity of Naber Konfeksion sh.p.k depends on the instructions of the head office for the output manufactured in the subsidiary.

- Valcuvia Alba sh.p.k is categorized as a truncated enclave with a moderate degree of integration in the host economy due to the absence of any linkages with suppliers or subcontractors, limitations in the training of the labor force but considerable investments already made and expected to be made. Referring to the complexity of activities realized in the host territory, its operations regard only assembly services resulting in little upgrading of this subsidiary in Albania. Regarding the relationship with the head office, Valcuvia Alba sh.p.k strictly follows all the instructions and guidelines of the head office in the overall operations of the subsidiary.

- Industria Ballkanike sh.p.k is also categorized as a truncated enclave as since starting of activity in Albania it has achieved only a small degree of integration in the host economy as it does not have any linkages with local suppliers or subcontractors, and qualification of the labor force remains only within its employees. However, with regard to investments it is one of the major investments made in the region and an important employment generator. This subsidiary in Albania serves to its customers only for services falling within production and only for cut, make, trim services. The relationship of Industria Ballkanike sh.p.k with the head office is categorized by a high degree of dependency in undertaking daily operations of the subsidiary as one of the owners of the group is present on a daily basis in the subsidiary.

Chapter 8
Innovation as a Prerequisite for Trade Openness in Emerging Economies

Abstract The majority of studies that focus on firm level data have found a positive causal effect of innovation to exports and vice versa. Both product and process innovation raise productivity that consequentially contribute to economic growth. Economists and researchers suggest that exporting firms become more innovative than those firms that operate only in the domestic market, because they gain experience in the foreign market. This chapter explores the impact of innovation on trade openness of emerging economies based on several models using OLS and data from international databases like EORA, ICIO and World Integrated Solutions. The results indicate that emerging economies need to improve their innovation capabilities and data collection process in order to achieve better results with regard to trade openness and integration into global value chains.

Keywords Innovation efficiency · Openness index · Knowledge creation · Market sophistication · Absorptive capacity

8.1 Introduction

In June 2011, Albania was a beneficiary of the agreement between the World Bank and the European Commission that provided technical assistance for the development of the Western Balkans Regional Research and Development Strategy for Innovation, called WBRIS-TA [52]. The strategy was implemented during the period 2011–2013, with the aim to strengthen the region's research capacity, enhance intraregional cooperation, promote collaboration with business sectors, explore possibilities for financing R&D from various funding sources, and help integrate the region into the European Research Area and Innovation Union [52]. In the last decade, Albania has made progress in improving the scores of indexes that evaluate institutions, market sophistication, knowledge and technology output, and creative output. Despite the slow growth, Albania ranked 83rd out of 126 countries in the Global Innovation Index 2018, leaving behind only North Macedonia out of WB countries.

© Springer Nature Switzerland AG 2020
J. Kacani, *A Data-Centric Approach to Breaking the FDI Trap Through Integration in Global Value Chains*, Lecture Notes on Data Engineering and Communications Technologies 50, https://doi.org/10.1007/978-3-030-43189-1_8

According to economic theory, innovation affects productivity, which consequentially contributes to economic growth. Innovation relates to the development and application of new ideas and technologies that improve the quality of goods and services, or make the process of production more efficient [1]. Thus, it is important to distinguish between product and process innovation. While there is a considerable experience accumulated in the field of innovation policy in developed countries, much of it is not directly applicable to developing countries [53]. In developed countries, researchers have found a positive causal effect between innovation and exports, where firms that invest in research and development activities, have a higher probability of exporting. These studies are usually conducted at firm level or sectorial level by using probit models or country level by using panel data. Tuhin [45] in his study of enterprise level panel data in Australia concludes that product innovation has a larger impact on inducing innovation in line with endogenous growth models. Also, Hausmann and Hidalgo [17], indicate that globally developed activities for generation of new products or identification of improved ways to produce existing products that are tradable across countries participating in GVCs.

The two most important channels through which developing countries can connect to the world economy and GVCs are through foreign direct investments and exports as they serve as channels for technology transfer [3]. In the economic models of innovation and growth, the increase of exports driven by innovation will in turn increase investments of the firm in activities stimulating innovation such as research and development activities, thereby raising the endogeneity issue between innovation and exports [50]. Based on this issue, the causal effect of the current innovation level in Albania on it's export performance aims to contribute to the existing literature on the effect of innovation on exports and vice versa. In the best of knowledge to the existing literature, there has not been any study on the direction of causal effects between innovation and exports in Albania. The analysis is based on constructing a model to find the direction of the causal effect between innovation and exports, after running several OLS models including as variables different proxies for innovation.

8.2   Innovation, International Trade, and Economic Growth

Many studies have accounted the importance of innovation on the economic growth. According to Pece et al. [30], innovation is a prerequisite for sustainable economic growth. They use a panel data set for the period 2000–2013 and find that economic development of Poland, Czech Republic and Hungary were influenced by innovation, research and development activities, human capital and foreign direct investment. Karaca [22] suggests that developing countries should invest in innovation in order to increase economic growth and catch up with developed countries. Verdier et al. [48] uses a panel data of 63 countries, which are mainly low-income and finds that innovation has a positive effect on productivity, which consequentially has a major influence on financial markets and facilitates the way to economic growth [39].

Wang et al. [51], Sanso et al. [36], Karaca [22], Pece et al. [30] use the number of patents as a proxy for innovation, because it serves as a measure for both product and process innovation. Smith and Thomas [42] use patent applications for new developed technologies at firm level in Russia. Triguero and Fernández [44] use panel data of Spanish manufacturing firms and find that technological collaboration with universities and external R&D are positively influencing innovation in Spain. On the other hand, Copenhagen Economics [6] argues that nations focused on technology sectors report a large number of patents, while nations based on agriculture generate less patents. Thus, according to them the economic structure of the nation or the region has a significant effect on the assimilation of process innovation. Moreover, alternative ways of explaining innovation, such as R&D expenditures [30], turnover or value added per employee [22], have also been used.

Piekkola [32] uses regional Finish firm level data and shows that R&D activity increases specialization in exports. Exports are necessary to improve external trade balance, as well as relate to productivity growth. Firm-level studies have found that innovation activity improves export performance, rather than vice versa [49]. Innovation output indicators, such as product and/or process innovations or patents, are found to positively affect export intensity and/or the probability of firms becoming exporters [32].

Investing in innovation would not only benefit firms at a national level but also at a regional level. López-Bazo and Motellón [26] find that the regions' variability significantly influences innovation. Among external factors, regional R&D expenditures have the highest impact in innovation, where an increase in the ratio of R&D to GDP of the region increases the probability of product innovation for the firms of the region. However, this effect depends on the firms' absorptive capacity of innovation, in both product and process innovation.

In addition, inward FDI can positively contribute to regional innovation, because of technological spillovers. Local firms gain knowledge not only from technological and products development of foreign firms, but also from foreign workers [4]. However, if the most skilled labor force is drawn by foreign firms, then the presence of FDI can make domestic firms experience losses in human capital and even threaten their existence [21]. The direction of these effects is based on a number of factors such as the type of investment, and the regional conditions to absorb new technologies [2, 8].

Wang et al. [51] find that in developing economies, FDI has a significant effect in the promotion of regional innovation and FDI spillovers are maximized by a more diverse type of industry rather than a specialized one, due to the fact that the last one is usually less flexible. Nevertheless, results show that FDI in a specialized industry has still a positive impact on innovation. Knowledge sharing is easier in this type of industry as it has lower costs [23, 38] and a higher concentration of R&D resources. Schäffler et al. [40] highlight the importance of human capital factor in attracting FDI in the case of German FDI's in Czech Republic. Also, Seker [41] in a panel data of enterprises in 43 developing countries concluded that enterprises which participate in international trade both as exporters and importers are more inclined to innovation than sole exporters or importers.

Moreover, Coad and Vezzani [5] and Fuchs [12] suggests that in case MNEs are able to have one large production facility in low-wage countries will gradually lead to knowledge transfer of high value added tasks (related to innovation, design and R&D). The knowledge transfer occurs from continuous interaction between workers and engineers inducing productivity improvements and process innovations [33]. The complexity of manufacturing processes and products can be approached as a reliable indicator of how well an economy can be expected to perform in the coming years [13, 18].

Smith and Thomas [42] find that FDI inflows and regional absorptive capacity have been the two key factors that boosted innovation in Russia during the period 1997–2011. Lin and Lin [25] and Hasanov et al. [16] find that inward FDI, outward FDI, and imports have a positively significant effect on innovative activities, while exports are statistically insignificant. They use firm level data from Taiwan Technological Innovation Survey for 1998–2000 period and construct a probit model. They argue that an increase in the level of competition in the domestic market will encourage firms to innovate in order to increase their efficiency and maintain their market status. Another study using 2005–2012 firm level panel data on Vietnam concludes that enterprises learn from exporting and the majority of gains is achieved if the enterprise is foreign owned, having the possibility to operate in a larger market [28]. In addition, Coad and Vezzani [5] and Delgado et al. [9] finds that exports are not statistically significant to affect innovation and argue that a large number of exporting firms have high productivity so it is not possible to further improve their level of technology or productivity through exports.

Another economic theory suggests that exports and innovation are complementary to the grow of an enterprise growth and implementing them both will increase its sales. An exporting enterprise has the opportunity to learn more from the foreign market and become more innovative than an enterprise that operates only in the domestic market. Mattoussi and Ayadi [27], Golovko and Valentini [14] use a fixed effect model and probit model to test the complementary nature of exports and innovation by using an unbalanced panel data of Spanish manufacturing enterprises for the period 1990–1999. They find that innovation and export are complementary activities, which leads to a significant positive effect to the growth of the enterprise. This occurs as exports facilitate participation in new markets where enterprises can enlarge their market share due to acquisition of new knowledge followed by improvements in the variety and quality of products. Similar results are found from Fuchs [12], Harris and Moffat [15] that identify spending on R&D in manufacturing firms as a determinant that improves the possibility for new products and services and facilitates creation of a knowledge pool that facilitates further integration in GVCs [24].

To continue, Saridakis et al. [37] uses a large data set from SMEs located in the United Kingdom and finds that innovative SMEs have a higher likelihood to export than non-innovative ones. They conclude that innovative products and services are profitable at a national level for the United Kingdom, because they improve the country's balance of payment.

Hasanov et al. [16] use an unbalanced panel data of 48 Asian countries with time series ranging from 1997 to 2011 and find that the number of patents, used as a proxy for innovativeness, is negatively affecting exports. On the other hand, R&D expenditure, trademarks, and publication of scientific journals was insignificant in affecting country's exports.

Triguero and Fernandez [44] discuss the importance of open innovation especially when there is collaboration between universities and firms. Open innovation means that valuable ideas come from inside or outside the company and can go to the market [47]. Openness boosts innovation intentionally through external knowledge sources and unintentionally through its spillovers. The positive influence of openness by knowledge spillovers affects not only firms that use the same technology but also those located near them. Elvekrok et al. [10], Newman et al. [29], Vannoorenberghe [47] and Etzkowitz and Leydesdorff [11] explain that triple helix networks are very efficient in promoting innovation. These networks consist of firms, universities, and governments that collaborate in a regional level. The benefits of such networks include: (i) risk sharing, (ii) skills and knowledge sharing, (iii) protecting property rights (iv) gaining access to new technologies and markets [34].

8.3 Innovation in the Western Balkans Countries

Innovation is commonly associated with developed countries. Nevertheless, developing countries such as Albania and other WB countries (Kosovo, Montenegro, North Macedonia, Serbia, Bosnia and Hercegovina) are trying to improve their innovation policies and catch up with developed economies.

According to the Cornell University, INSEAD, WIPO [7], Albania together with other WB countries have improved some determinants associated with innovation such as scores on indexes referring to knowledge creation, knowledge absorption, and knowledge impact. However, there are several problematic issues such as the fact that Albania's index score (see Fig. 8.1) for knowledge creation ranges from 0 to 5 throughout the years, while the score's boundaries are from 0 to 100. Knowledge creation index captures the situation of patents, scientific articles, and citable documents in the sub index of Cornell University, INSEAD, WIPO [7]. Albania's knowledge absorption index score has risen, reaching the highest value in 2015 due to the rise of FDI net inflows as percentage of GDP. However, in 2016 it decreased again due to the decrease of high tech net imports as percentage of GDP and the decrease of FDIs. Human capital and research index score has been declining according to the Cornell University, INSEAD, WIPO [7]. In 2015, one of the indicators of human capital and index score related to the number of researchers per million of population decreased to 147.9 from 545.2 in 2014. Also, in 2017, a significant decrease occurred in another indicator of the index related to the growth rate per worker. A slight increase is observed in the number of ISO 9001 quality certificates [23].

The current degree of innovation in WB countries is affected from the positive economic development during the last decade. The WB countries experienced positive growth rates and declining poverty rates [53]. With reference to Table 8.1 in 2018, Montenegro had the highest economic growth rate of 4.9%, while North Macedonia

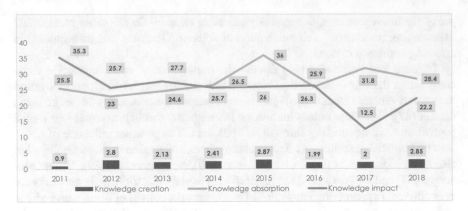

Fig. 8.1 Knowledge absorption, knowledge creation, and knowledge impact in Albania (2011–2018). *Source* Global Innovation Index report, 2011–2018

Table 8.1 General economic information about countries in the Western Balkans (2018)

	Albania	Bosnia and Herzegovina	Kosovo	Montenegro	North Macedonia	Serbia
Population	2,866,376	3,323,929	1,845,300.00	622,345.00	2,082,958.00	6,982,084.00
GDP in mln EUR	12,308	16,765	6,695	4,620	10,739	42,805
GDP per capita in EUR	4,294.00	5,043.75	3,628.40	7,424.69	5,155.95	6,130.81
Growth rate (%)	4	3.1	4.1	4.9	2.7	4.3
Inflation (%)	2	1.2	1.5	2.6	1.5	2.0
Unemployment rate	12.83	18.40	29.5	15.18	19.4	13.3

Source World Bank and National Statistical Institute of each country

had the lowest economic growth rate of the region of 2.7%. The economic growth is attributed to the rise of investments and consumption because of the increase in wages and pensions [19]. Furthermore, the region achieved to attract the highest number of green-field FDI projects in the past six years. WB countries are enhancing their position in the global market, as their exported goods have become more sophisticated resulting in the increase in the level of exports with European Union the main trade partner of the WB countries. Trade with the European Union has a total value of imports of EUR 22.6 billion and a total value of exports of EUR 31.8 billion [53].

To continue, Table 8.2 shows the overall situation of innovation in the WB countries (except Kosovo). In 2018, Montenegro achieved the highest rank in the region leaving behind Serbia, Bosnia and Hercegovina, Albania, and North Macedonia. According to GII Report (2018), Montenegro is described as an innovation achiever due to the good performance in innovation in comparison to the other countries with similar level of development [20].

Table 8.2 Main determinants of the Global Innovation Index for the countries in the Western Balkans (2018)

	Albania	Bosnia and Herzegovina	Montenegro	Serbia	North Macedonia
Rank from 126 countries	83	77	52	55	84
Global Innovation Index	30	31.1	36.5	35.5	29.9
Innovation efficiency ratio	0.4	0.5	0.6	0.6	0.5
Innovation input sub-index	41.6	41.6	44.8	43.5	40.7
Innovation output sub-index	18.4	20.6	28.2	27.4	19.1

Source Global Innovation Index Report, 2018

Another important determinant of the performance on innovation for a country is the innovation efficiency ratio ranging from 0 to 1. This ratio shows how efficient is the use of innovation inputs for the production of innovation outputs. Albania has the lowest innovation efficiency ratio compared to the other WB countries. Albania's number of researchers per million of population and Gross Expenditure in Research and Development (GERD) as percent of GDP has remained constant at 0.2 throughout the years. Moreover, Albania's GERD as percentage of GDP is low compared to EU's, which is 2.07%. Albania and the other WB countries need to increase their GERD as percentage of GDP in order to establish an efficient system for R&D, which will boost the innovation systems of the region [43].

Albania was featured in Cornell University, INSEAD, and WIPO [7] as one of the middle income countries which over performed the high-income ones regarding market sophistication (see Table 8.3). WB countries scores for innovation inputs such as institutions and market sophistication are higher than scores for innovation output, due to an increase in the ICT industry an increase in investments, and ease of getting credit [20]. On the other hand, scores of innovation outputs related with creativity, knowledge and technology are below the average due to the low number of patents, scientific and technical articles, citable documents and creative contents. In order to increase efficiency ratios, WB countries should be more focused in improving the quality of innovation outputs.

Another factor that is contributing to the low level of human capital and research is the "brain drain" phenomenon, which is the emigration of qualified people. The migration of young qualified people is a common trend throughout the countries of WB and needs to be addressed because it influences several aspects of the business environment such as competitiveness, innovation, and growth [53]. According to United Nations Industrial Development [46] report, the reversing of the brain drain phenomenon can be done by designing policies to return Ph.D. graduates and attract

Table 8.3 Sub-indexes of Global Innovation Index report for the countries in the Western Balkans (2018)

	Albania	Bosnia and Herzegovina	Montenegro	Serbia	North Macedonia
Institutions	64.5	58.7	68.2	67.2	67.5
Human capital and research	21.3	41.4	33.4	32.2	27.3
Infrastructure	45.3	34.5	47.8	49.6	39.6
Market sophistication	52.3	43	42.8	39.2	45.2
Business sophistication	24.4	30.4	31.6	29.2	24.2
Knowledge and technology outputs	13.7	20.2	16.3	26.7	20.9
Creative outputs	23.1	21	40.2	28.1	17.3

Source Global Innovation Index Report, 2018

the Albanian Scientific Diaspora. There are some policies that need to be implemented in order to have economic growth, political stability, low levels of corruption, and innovative universities and research institutes. The second way of reversing the "brain drain" phenomenon is to mobilize the Diaspora through collaboration with universities and research institutions in Albania.

8.4   The Interrelation of Trade Openness and Innovation in Albania

This section investigates at the country level the relationship between openness and innovation inputs and outputs for Albania. To the best knowledge so far similar studies are not available in the existing literature. One of the constrains of this study is that there is a small number of observations for innovation, because Albania as a developing country has only a few yearly observations starting after the 2000s.

Data description

The main source for innovation data is the Global II Report (2010–2018), which is published every year starting from 2008. Albania was included in the report starting from 2010. This study uses yearly observations for the period 2010 to 2018 for the innovation inputs and outputs variables.

The main dependent variable is the Openness Index, which is an economic metric calculated as the ratio of country's total trade (sum of exports and imports), to the country's gross domestic product. The interpretation of the Openness Index is that the higher the index the larger the influence of trade on domestic activities, and the stronger the country's economy. The Openness Index consists of yearly observations for the period 2010–2018 as well, the data available on innovation variables for Albania.

In addition, with reference to previous studies macroeconomic indicators serving as control variables such as FDI inflow to GDP ratio, GDP per capita, unemployment rate are included in the estimation of trade openness innovation models in Albania.

Table 8.4 provides the description, explanation, and the expected sign of independent variables. The expected sign refers to the impact the variable might have in case a significant causal effect is observed on our dependent variable, Openness Index.

Initially, for all dependent and independent variables the descriptive statistics are analysed. Table 8.7 placed in the appendix of this chapter includes a summary of descriptive statistics of all the variables included in the estimated models.

Since there is a small number of observations the OLS (Ordinary Least Squares) model is used to check if innovation variables have any causal effect on the dependent variable Openness Index. The model is defined the following way:

$$Y_t = \alpha + \beta X_t + \varepsilon_t$$

- Y_t refers to the dependent variable at time t;
- α is the intercept;
- X_t is the vector of independent variables at time t
- ε_t is the error term.

In order to achieve the best model the following steps are necessary. Firstly, the descriptive statistics of the variables are analyzed. After this analysis the Dicky-Fuller test is performed to check if the variables are stationary (see Table 8.5). A stationary time series is one whose statistical properties such as mean, variance, and autocorrelation are constant over time. It is important to have stationary variables, because most statistical forecasting methods are based on the assumption that time series can be rendered approximately stationary (i.e., "stationarized") through the use of mathematical transformations.

Dicky-Fuller Test for stationarity:
H_0: variable is non-stationary (e.g. there exists a unit root in the variable)
H_A: variable is stationary (e.g. there doesn't exist any unit root in the variable)
Decision Criteria: Reject the null if p-value is less than 1%, 5%, or 10% level of significance.

What follows is a correlation analysis between the independent variables. The decision criteria is that if the correlation between two variables is more than 0.8 in absolute value, those two variables will not be considered in the same regression model. The correlation matrix of the independent variables is presented in Table 8.8 in the appendix of this chapter. The variables that present problematic correlation are eliminated from regressions in order to find the best fitted model.

Table 8.4 The explanation, description and the expected sign of variables used in the estimated models

Variable Name	Description	Explanation	Expected sign
FDI_GDP	FDI inflow to GDP ratio. *Source* Bank of Albania	The ratio of FDI inflow to GDP is considered a control variable. There are several research papers that have found that FDI inflow to GDP ratio has a positive significant impact on exports [35], as well as a positive significant effect on innovation [51]	+
lGDP_cap	GDP per capita is the measure of the total output of a country where the Gross Domestic Product (GDP) is divided by the total population in the country. In order to capture the percentage change (elasticity) rather than the unit change, the GDP per capita (USD) is transformed to its logarithmic value. GDP per capita is a widely used measure of economic activity. *Source* World Bank	The logarithm of GDP per capita is considered as a control variable. GDP per capita is a proxy for the financial resources of the state [31], as well as an estimate of the size of technological development of the country	+
Unemp_rate	The unemployment rate is the number in the civilian labor force divided by the number of unemployed. *Source* Bank of Albania	Unemployment rate is considered as a control variable, assumed to be negatively related to innovation and exports [30, 31]	−
TertED	Tertiary Education is a sub-index of Human Capital and Research, and is normalized in a range from 0 to 100. It includes: tertiary enrolment, graduates in science and engineering, tertiary inbound level mobility. *Source* Global Innovation Index Report	Tertiary education is an innovation input. The more educated the population, the highest the innovation score for the country. It is expected to have a positive effect on exports	+

(continued)

Table 8.4 (continued)

Variable Name	Description	Explanation	Expected sign
RD	Research and Development is a sub-index of Human Capital and Research, and is normalized in a range from 0 to 100. It includes number of researchers per million of population, GERD (Gross Expenditure on Research and Development), global R&D companies, QS rankings. *Source* Global Innovation Index Report	Research and Development is used as a proxy for innovation input [16, 30–32, 44]	+
Know_work	Knowledge workers index score is reported in GII Report and includes: Employment in knowledge-intensive services; Firms offering formal training; GERD as, %GDP performed by business enterprise; GERD financed by business enterprise (% of total GERD); females employed with advanced degrees, % total employed. This variable is used as a proxy for innovation. *Source* Global Innovation Index Report	Gross expenditure on research and development as percentage of GDP has been used as a proxy for innovation by many researchers [26]	+
Innov_link	Innovation Linkages score is reported in Global Innovation Index Report and includes: University/Industry research collaboration; State of cluster development; GERD financed by abroad (% of total GERD). This variable is used as a proxy for innovation. *Source* Global Innovation Index Report	University/Industry research collaboration has been used by many researchers as a proxy for innovation. Triguero and Fernandez, [44] find that the higher the knowledge sharing with universities the higher the probability of innovation	+
Know_abs	Knowledge absorption score is reported in GIIReport and includes: intellectual property payment, high-tech net imports as % of total trade, ICT services import as % total trade, research talent in business enterprise, FDI net inflows. *Source* Global Innovation Index Report	Knowledge absorption is an innovation input	+

(continued)

Table 8.4 (continued)

Variable Name	Description	Explanation	Expected sign
Know_cre	Knowledge creation score is reported in the GII and includes: patent applications by origin; PCT international applications by origin. This variable is used as a proxy for innovation. *Source* Global Innovation Index Report	Knowledge creation is an innovation output. The number of patents has been commonly used as a proxy for innovation, especially for firm level studies in developed countries [16, 30, 31]	+
Know_imp	Knowledge impact score is reported in the GIIReport and includes: growth rate of GDP per person engaged; new business density; total computer software spending %GDP; ISO 9001 quality certificates; high-tech and medium high-tech output, % total manufactures output. This variable is used as a proxy for innovation. *Source* Global Innovation Index Report	Knowledge impact is an innovation output and it is expected to have a positive impact	+

Table 8.5 Variables stationarity resulting from Dicky-Fuller test on innovation trade openness index models in Albania

Variable name	Value	Value at first differencing	Value at second differencing
Openness	Non-stationary	Stationary*	
FDI_GDP	Non-stationary	Stationary***	
lGDP_cap	Non-stationary	Non-stationary	Stationary*
Unemp_rate	Stationary**		
TertRD	Stationary*		
RD	Stationary***		
Know_work	Non-stationary	Stationary*	
Innov_link	Non-stationary	Stationary***	
Know_abs	Non-stationary	Stationary***	
Know_cre	Stationary***		
Know_imp	Non-stationary	Stationary***	

Significant at ***1%, **5%, *10%

Next, the Breusch–Godfrey serial correlation LM test is applied. This is a test for autocorrelation in the errors of the regression model and it makes use of the residuals from the model , which are considered in a regression analysis.

H_0: there is no serial correlation
H_A: there is serial correlation
Decision Criteria: Reject the null if p-value is less than 1%, 5%, or 10% level of significance. Serial correlation is defined as the correlation between the observations of residuals and it is commonly found in time series data.

Finally the Breusch-Pagan Test for heteroscedasticity is applied. This test is usually applied by assuming that heteroskedasticity may be a linear function of all the independent variables in the model, which is a weakness of this test.

H_0: there exists homoscedasticity (e.g. error variances are all equal)
H_A: there exists heteroscedasticity (e.g. error variances are a multiplicative function of one or more variables)
Decision Criteria: Reject the null if p-value is less than 1%, 5%, or 10% level of significance.

The application of the above mentioned tests and the considerable number of trials undertaken the three best models based on OLS estimations are presented below. These models are selected models based on the highest R-squared. It is a statistical measure that represents the proportion of the variance for a dependent variable that's explained by an independent variable, as well as higher number of significant coefficients.

1.   openess $= \alpha + \beta_1 \text{TertED} + \beta_2 \text{RD} + \beta_3 \text{Know\_work} + \beta_4 \text{Innov\_Link} + \varepsilon$
2.   openess $= \alpha + \beta_1 \text{lGDP\_cap} + \beta_2 \text{TertED} + \beta_3 \text{RD} + \beta_4 \text{Know\_cre} + \varepsilon$
3.   openess $= \alpha + \beta_1 \text{FDI\_GDP} + \beta_2 \text{unemp} + \beta_3 \text{Know\_abs} + \beta_4 \text{Know\_imp} + \varepsilon$

8.5   Results

Table 8.6 summarizes the results from the OLS estimations and the level of significance of the independent variables in the causal effect they have on the Openness Index.

The results from OLS estimations indicate that the models do not show robust results on casual effects. With reference to model 1 and model 2 the R-squared in very high even though the there are some significant variables, while model 3 has resulted in only insignificant variables. Model 1, captures a significant positive effect of tertiary education and R&D to Openness Index. Thus the higher tertiary enrolment the higher the Openness Index of the country. Moreover, R&D has a positive significant effect, which is intuitive because the higher the investment in R&D the higher the level of Openness Index for the country.

The tests for serial correlation and heteroscedasticity are not relevant in this case, as the models are not robustly significant. These results can be explained due to the lack of data, and small number of observations for innovation. Moreover, despite that the results of the models do not show robust results it is not possible assert the

Table 8.6 Results from OLS estimations for our three best models on innovation trade openness index models in Albania

Independent variable	Model 1	Model 2	Model 3
D. FDI_GDP			−184.3825
D2. lGDP_cap		0.0003197	
unemp			−0.0055786
D. TertRD	0.0160295***	−0.0025604	
D.RD	0.0027591**	−0.0003152	
D. Know_work	0.0009178		
D. Innov_link	0.0012624		
D. Know_abs			−0.0008761
Know_cre		−0.0846121*	
D. Know_imp			−0.0012834
Constant	0.0038077	0.202853*	−0.0072637
R-squared	0.9968	0.9274	0.2425

Significant at ***1%, **5%, *10%

impact of additional indicators of innovation such as human capital and research, research and development, knowledge workers, knowledge absorption, innovation linkages, knowledge creation, knowledge diffusion are not directly influencing a country's openness. Our findings are also supported by other researchers. Hasanov et al. [16] identified that R&D expenditure, trademarks, and publication of scientific journals were insignificant in affecting Asian countries' openness, especially exports. Coad and Vezzani [5], also find no robust relationship between trade and exports in manufacturing. More significant results can be obtained in the future by collecting industry level data, and more specifically firm level data in order to see how innovation affects the level of exports and openness to the markets.

Appendix

See Tables 8.7 and 8.8.

Table 8.7 Descriptive statistics of the equation variables

Variable	Observations	Mean	Standard deviation	Min.	Max.
openness	9	0.7909684	0.0284952	0.7426498	0.836036
FDI_GDP	9	0.0000842	0.0000101	0.000068	0.0000991
GDP_cap	9	4496.444	323.3528	4094	5075
unemp_rate	9	14.74319	1.773061	12.34	17.49
terted	8	28.7375	4.503629	23.8	37.1
rd	8	5.46625	8.071342	1.08	24.4
know_work	8	24.65	3.919184	19	30.3
innov_link	8	17.475	3.70704	10.5	22.2
know_abs	8	27.7125	4.266292	23	36
know_cre	8	2.24375	0.6589155	0.9	2.87
know_imp	8	25.175	6.334655	12.5	35.3

Table 8.8 Correlation matrix of the independent variables

	lfdiin~d	unemp_~e	terted	rd	know_w~k	innov_~k	know_abs	know_cre	know_imp
lfdiinflow~d	1.0000								
unemp_rate	0.0247	1.0000							
terted	0.6502	0.4180	1.0000						
rd	−0.6593	−0.3406	−0.5204	1.0000					
know_work	0.2423	0.4820	0.5163	−0.4258	1.0000				
innov_link	0.0939	−0.2479	−0.4991	−0.3771	−0.3146	1.0000			
know_abs	0.0558	0.2712	−0.2020	−0.3680	−0.3417	0.5938	1.0000		
know_cre	0.1978	0.1608	−0.0526	−0.7094	0.4549	0.4721	0.2833	1.0000	
know_imp	−0.4556	0.1560	−0.1912	0.6768	−0.1258	−0.6228	−0.4357	−0.4106	1.0000

References

1. Antonietti R, Bronzini R, Cainelli G (2017) Inward greenfield FDI and innovation. J Ind Bus Econ 42(1):93–116. https://doi.org/10.1007/s40812-014-0007-9
2. Armas E, Rodríguez J (2017) Foreign direct investment and technology spillovers in Mexico: 20 years of NAFTA. J Technol Manag Innov 12(3):34–47. https://doi.org/10.4067/s0718-27242017000300004
3. Aw B, Roberts M, Yi XD (2011) R&D investments, exporting, and productivity dynamics. Am Econ Rev 101(4):1312–1344
4. Buckley PJ, Clegg J, Wang C (2010) Is the relationship between inward FDI and spillover effects linear? An empirical examination of the case of China. In Foreign direct investment, China and the world economy. Palgrave Macmillan, London, pp 192–215
5. Coad A, Vezzani A (2017) Manufacturing the future: is the manufacturing sector a driver of R&D, exports and productivity growth? JRC working papers on corporate R&D and innovation, No 06/2017. Joint Research Centre, Spain
6. Copenhagen Economics (2016) Towards a Foreign Direct Investment (FDI) attractiveness scoreboard. Directorate General for Internal Market, Industry, Entrepreneurship and SMEs Unit 01 Economic Analysis, European Commission
7. Cornell University, INSEAD, WIPO (2018) The global innovation index 2018: energizing the world with innovation. Ithaca, Fontainebleau, Geneva
8. Crespo N, Fontoura MP (2007) Determinant factors of FDI spillovers-what do we really know? World Dev 35(3):410–425
9. Delgado MA, Farinas JC, Ruano S (2002) Firm productivity and export markets: a nonparametric approach. J Int Econ 57(2):397–422
10. Elvekrok I, Veflen N, Nilsen ER, Gausdal AH (2018) Firm innovation benefits from regional triple-helix networks. Reg Stud 52(9):1214–1224
11. Etzkowitz H, Leydesdorff L (2000) The dynamics of innovation: from national systems and "Mode 2" to a triple helix of university–industry–government relations. Res Policy 29(2):109–123
12. Fuchs ER (2014) Global manufacturing and the future of technology. Science 345(6196):519–520
13. Gelebo E, Plekhanov A, Silve F (2015) Determinants of frontier innovation and technology adoption: cross-country evidence. Working Paper No. 173, European Bank for Reconstruction and Development, United Kingdom
14. Golovko E, Valentini G (2011) Exploring the complementarity between innovation and export for SMEs' growth. J Int Bus Stud 42(3):362–380
15. Harris R, Moffat J (2011) R&D, innovation and exporting. UK Spatial Economics Research Centre, Discussion Paper No. 73, United Kingdom
16. Hasanov Z, Abada O, Aktamov Sh (2015) Impact of innovativeness of the country on export performance: evidence from Asian countries. J Bus Manag 17(1):33–41
17. Hausmann R, Hidalgo CA (2011) The network structure of economic output. J Econ Growth 16(4):309–342
18. Hidalgo CA, Hausmann R (2009) The building blocks of economic complexity. Proc Natl Acad Sci 106(26):10570–10575
19. Hu AG, Jefferson GH, Jinchang Q (2005) R&D and technology transfer: firm-level evidence from Chinese industry. Rev Econ Stat 87(4):780–786
20. Jacobs J (1969) The economy of cities. Random House, New York, NY
21. Kacani J, Van Wunnik L (2017) Using upgrading strategy and analytics to provide agility to clothing manufacturing subsidiaries: with a case study. Glob J Flex Syst Manag 18(1):21–31
22. Karaca Z (2018) Research, innovation, and productivity: an econometric analysis at the manufacturing sector in Turkey. Cumhuriyet Üniversitesi İktisadi ve İdari Bilimler Dergisi 19(2):44–56
23. Khachoo Q, Sharma R (2016) FDI and innovation: an investigation into intra- and inter-industry effects. Glob Econ Rev 45(4):311–330. https://doi.org/10.1080/1226508x.2016.1218294

24. Leman E, Göçer I (2015) The effects of foreign direct investment on R&D and innovations: panel data analysis for developing asian countries. Soc Behav Sci 195:749–758
25. Lin HL, Lin ES (2010) FDI, trade, and product innovation: theory and evidence. South Econ J 77(2):434–464
26. López-Bazo E, Motellón E (2018) Innovation, heterogeneous firms, and the region: evidence from Spain. Reg Stud 52(5):673–687
27. Mattoussi W, Ayadi M (2017) The dynamics of exporting and innovation: evidence from the Tunisian manufacturing sector. J Afr Econ 26(1):52–66
28. Newman C, Page J, Rand J, Shimeles A, Söderbom M, Tarp F (2016) Made in Africa: learning to compete in industry. Brookings Institution Press, Washington, DC
29. Newman C, Rand J, Tarp F, Thi Tue Anh N (2017) Exporting and productivity: learning from Vietnam. J Afr Econ 26(1):67–92
30. Pece AM, Simona OEO, Salisteanu F (2015) Innovation and economic growth: an empirical analysis for CEE countries. Procedia Econ Financ 26:461–467
31. Petrariu IR, Bumbac R, Ciobanu R (2013) Innovation: a path to competitiveness and economic growth. The case of CEE countries. Theor Appl Econ 20(5):15–26
32. Piekkola H (2018) Internationalization via export growth and specialization in finnish regions. Cogent Econ Financ 6(1):1514574. https://doi.org/10.1080/23322039.2018.1514574
33. Pisano G, Shih W (2012) Producing prosperity: why America needs a manufacturing renaissance. Harvard Business Review Press
34. Pittaway L, Robertson M, Munir K, Denyer D, Neely A (2004) Networking and innovation: a systematic review of the evidence. Int J Manag Rev 5(3–4):137–168
35. Prasanna N (2010) Impact of foreign direct investment on export performance in India. J Soc Sci 24(1):65–71. https://doi.org/10.1080/09718923.2010.11892838
36. Sanso-Navarro M, Vera-Cabello M (2018) The long-run relationship between R&D and regional knowledge: the case of France, Germany, Italy and Spain. Reg Stud 52(5):619–631
37. Saridakis G, Idris B, Hansen JM, Dana LP (2019) SMEs' Internationalization: when does innovation matter? J Bus Res 96:250–263
38. Saxenian A, Hsu JY (2001) The Silicon Valley-Hsinchu connection: technical communities and industrial upgrading. Ind Corp Change 10(4):893–920
39. Schumpeter J (1942) Creative destruction. Capitalism, Socialism Democracy 825:82–85
40. Schäffler J, Hecht V, Moritz M (2017) Regional determinants of German FDI in the Czech Republic: new evidence on the role of border regions. Reg Stud 51(9):1399–1411
41. Seker M (2011) Importing, exporting, and innovation in developing countries. Policy research working paper 5156, World Bank, Washington D.C.
42. Smith N, Thomas E (2017) Regional conditions and innovation in Russia: the impact of foreign direct investment and absorptive capacity. Reg Stud 51(9):1412–1428
43. Svarc J (2014) A triple helix systems approach to strengthening the innovation potential of the Western Balkan Countries. Int J Trans Innov Syst Intersci Enterprises 3(3):169–188
44. Triguero A, Fernández S (2018) Determining the effects of open innovation: the role of knowledge and geographical spillovers. Reg Stud 52(5):632–644
45. Tuhin R (2016) Modelling the relationship between innovation and exporting: evidence from Australian SMEs. Research paper 3/2016, Office of the Chief Economist, Department of Industry, Innovation and Science, Australian Government
46. United Nations Industrial Development (2018) Global value chains and industrial development. Lessons from china, south-east and south asian. United Nation Publication, Geneva, Switzerland
47. Vannoorenberghe G (2015) Exports and innovation in emerging economies: firm-level evidence from South-Africa. DFID Working Paper, Tilburg: Tilburg University
48. Verdier G, Kersting E, Dabla-Norris ME (2010) Firm productivity, innovation and financial development No. 10–49. International Monetary Fund, Washington D.C.
49. Wagner J (2012) International trade and firm performance: a survey of empirical studies since 2006. Rev World Econ 148(2):235–267

50. Wagner R, Zahler A (2015) New exports from emerging markets: do followers benefit from pioneers? J Dev Econ 114(C):203–223
51. Wang Y, Ning L, Li J, Prevezer M (2016) Foreign direct investment spillovers and the geography of innovation in Chinese regions: the role of regional industrial specialization and diversity. Reg Stud 50(5):805–822
52. World Bank (2013) Western Balkans regional R&D. Strategy for innovation under the Western Balkans Regional R&D Strategy for Innovation World Bank Technical Assistance Project, under financing of European Commission DG ENLARG-TF011064)
53. World Bank (2019) Rising uncertainties Western Balkans, Regular economic report No. 16, Washington D.C.

Chapter 9
The Road Ahead for Active Integration into Global Value Chains

Abstract This chapter presents the main findings of the research drawn from analyzing at a country, clothing industry, and enterprise level the integration of emerging economies into global value chains resulting from the presence of foreign direct investments. The chapter starts with conclusions referring to the clothing industry, the knowledge transferred in the host territory, and the evolution in the activity of clothing manufacturing subsidiaries operating in emerging economies. In addition, the chapter presents policy recommendations on the positive impact foreign investments can make in knowledge transfer and innovation capacities of developing countries to turn them into active players in global value chains.

Keywords Development policies · FDI trap · Clothing industry · Global value chains · Business environment

9.1 Breaking the FDI Trap and Integration into GVCs

Clothing industry in Albania needs to break the vicious circle of FDI and encourage a virtuous circle. This section tries to generalize on the progress made to break the vicious circle and move towards a virtuous one by focusing on the evolution of subsidiaries and industrialization in the host territory.

9.1.1 Breaking the FDI Trap—Evolution of the Subsidiary in the Host Territory

In order for Albania to break the FDI trap, it is critical to attract more quality FDI and to be part of a virtuous circle. A virtuous circle can occur if enterprises in Albania including those in the clothing industry upgrade. Upgrading in the clothing industry should not be limited only to foreign firms but it needs to include also local enterprises operating in this industry.

Clothing manufacturing enterprises need to obtain process, product, and functional upgrading. The four clothing manufacturing subsidiaries managed to achieve

© Springer Nature Switzerland AG 2020

J. Kacani, *A Data-Centric Approach to Breaking the FDI Trap Through Integration in Global Value Chains*, Lecture Notes on Data Engineering and Communications Technologies 50, https://doi.org/10.1007/978-3-030-43189-1_9

process upgrading mostly from the increase and diversity in the number of services performed in the host territory (starting from only sewing services to packing, embroidery, stamping, packing, and delivery). Product upgrading occurs mainly in the manufacturing of more complex standardized products and little in the manufacturing of differentiated products. Functional upgrading has been very limited in the clothing manufacturing subsidiaries. There is a need for services that can facilitate process and product upgrading. This requires the establishment of design and research centers that can help in local upgrading. Specialized testing, certification, and measurement can induce upgrading. Knowledge transferred in the host territory helps breaking the vicious circle and opens the road to the virtuous circle. To start with, the clothing manufacturing subsidiaries have been able to increase employment in Albania; however, the increase is dominated mostly by low skilled labor that are hired to perform assembly services. On average 5% of total employees are placed in high management, 10% fall into middle level management, and the remaining 85% are employees that provide low skilled services. The quality of local employees has not improved much as in three out of four subsidiaries labor tasks mostly include only assembly services (cut, make, and trim) by using the necessary equipment required to complete these tasks. Most local employees do not possess any previous experience or training even on basic assembly tasks. However, only a limited number of shop floor workers have become more specialized as they have been able to move within various departments in the subsidiary, like those that started with sewing, went into buttoning, than embroidery, etc. Local employees serving in middle and top management of the subsidiaries are more qualified; however, the number is small as few local employees are appointed in these positions. Local employees in management positions are better qualified as they have been able to gain both technical knowledge through various trainings or study visits in the head office and managerial knowledge due to frequent interactions with managers and administrators in the head office. They have also gained more knowledge from interaction with customers and suppliers compared to local employees allocated only to assembly services. On the other hand, local employees in production have mostly benefited from technical knowledge accumulated during the years they were exposed to new processes. Local employees mostly exposed to new processes are the ones that have been with the subsidiary for a number of years by getting involved in the quality control of raw materials, preparation of models, embroidery, etc. Despite that technical and managerial knowledge are within the clothing industry, a limited number of local employees have achieved a certain degree of professional progress as they have gained knowledge not previously available in the host territory and have mastered skills and that may be useful in learning similar tasks in other industries.

To continue, clothing manufacturing subsidiaries have weak linkages with the local economy resulting mostly from the absence of local firms that might serve as suppliers of clothing manufacturing subsidiaries even for simple raw materials like card boxes and plastic bags. Local suppliers are not able to meet the standards imposed by international customers of subsidiaries. Mixed evidence is drawn on linkages of subsidiaries with regard to subcontractors engaged in production. These linkages range from dependent in nature when the operational activity of subcontractors depends at a large extend on the orders it receives from the subsidiary as in the case of Shqiperia Trikot sh.p.k and independent linkages as in the case of Naber Konfeksion sh.p.k. Demonstration effects occurring from clothing manufacturing subsidiaries in the host territory are present only partially. Demonstration effects have occurred through opening up of local clothing manufacturing enterprises from local employees that previously served in middle management of subsidiaries. These newly created clothing manufacturing enterprises serve either as subcontractors to the subsidiary in which the owner previously worked for or as direct subcontractors to international customers. A good example are the former subcontractors of Naber Moden that initially offered services to international customers only through Naber Moden, while since 2015 have become independent of Naber Moden and are offering services directly to international customers.

9.1.2 Breaking the FDI Trap—Industrialization

The case study analysis indicates that even though the four clothing manufacturing subsidiaries operate in the same host territory, are part of the same industry, produce mostly standardized products, and are located in regions that are more or less homogenous, they have achieved different industrialization paths. A key determinant of the different evolution paths is the relationship of the subsidiary with the head office. The four clothing manufacturing subsidiaries started as receptive subsidiaries by depending for their operations entirely on the head office. With the passing of time, a relationship based on trust in local management transformed Shqiperia Trikot sh.p.k into an active subsidiary. Shqiperia Trikot sh.p.k has acquired from the head office a good degree of freedom in performing daily operations, in proposing new expansions projects, in undertaking more complex functions, in participating in the strategic meetings organized in the head office, and in setting business development strategies. The other three clothing manufacturing subsidiaries have maintained the dependency from the head office in the daily operations of the subsidiary. This dependency has not allowed these subsidiaries to go beyond provision of assembly services. The subsidiaries administered by foreign staff are the ones that have evolved the least. Despite of the different evolution paths the clothing manufacturing subsidiaries did not become better than the group as their strategy towards the host territory is mostly based on profit maximization.

Realization of more complex functions and products would strengthen location advantages of Albania by better serving customers through provision of more services and by reducing even further the delivery time of finished articles. The latter is of high importance at times when the clothing industry is heavily dominated by the concept of "quick fashion". A change in location advantages would favor clothing manufacturing enterprises to connect to key players of the value chain of the clothing industry opening up the opportunity to focus on complex operations like design, research and development, marketing and outsource services related to production to countries of North Africa like Tunisia and Morocco. A similar scenario occurred with South Korea and Taiwan that outsourced production in Bangladesh or Central America while retaining key operations in the host territory. The cheap labor force due to lower wages compared to neighboring countries and the rising costs encountered in Asian countries, are temporary location advantages that appeal to foreign clothing manufacturing to transfer production in Albania. However, in the near future production may be transferred in other host territories like African countries, where the labor force could be cheaper than in Albania, leaving the existing host territory without inheriting any qualitative effects.

Even though the clothing industry is labor rather than capital intensive, it remains an interesting industry for Albania in terms of obtaining industrialization. When this industry is considered as the first step of the ladder of industrialization, it mostly refers to assembly services of the industry and its basic tasks that may be replicated in other industries. This way the host territory may attract foreign investors in industries which are currently absent from Albania like toy, mobile, etc.

As introduced in Chap. 4, the low barriers to entry in the production segment of value chain of the clothing industry are accompanied with higher competition from a number of developing countries, making it more difficult to develop.

However, Albania can move up in the value chain of the clothing industry by revitalizing production of textiles that was fully operational before the 1990s a segment that is more capital intensive than clothing manufacturing. Availability in the host territory of fabrics that meet the criteria to be used in production of cloths would diminish imports of raw materials and would further reduce the delivery time to end customers, especially to those in the EU market. To make functional the textile industry in Albania is costly. It requires the necessary knowledge to generate quality output making it difficult for local enterprises that struggle to meet the standards even for basic raw materials like plastic bags. As such, foreign enterprises able to afford high capital investments and possess the knowledge on the textile industry are able to take this initiative. Participation of subsidiaries in such ventures can increase the benefits of the host territory. This is the case of Cotonella S.p.A and its subsidiary Shqiperia Trikot sh.p.k that are committed to establish a regional center for fabric and dyeing that is expected to put Albania into new segments of the value chain. In addition, the presence of the textile industry would facilitate creation of a base of local suppliers of raw materials like cotton and silk that have appropriate harvesting conditions in Albania. Long term advantages can be established in a country by encouraging global suppliers. There can be a reliance by local firms from global suppliers for a wide range of inputs and services. This can provide access to capabilities and up-to-date products and services when they participate in GVCs from the beginning and can lower the barriers and risks to entry for local firms.

A threat to industrialization of the host territory are factors that limit upgrading to which all the clothing manufacturing subsidiaries are exposed despite of being located in different regions. These factors include: (i) unstable electricity supply, (ii) lengthy custom procedures, (iii) frequent changes in the fiscal regime, and (iv) absence of economic zones. Another, threat is the small size of the population which means that as the number of clothing manufacturers in Albania goes up so will the competition to have access to more trained employees (middle management) in particular. Even though the unemployment rate in Albania is on average 16% of the labor force it is unlikely to favor employment in the clothing industry as the majority of the labor force has been to graduate schools due to the increase in the number of private universities that offer higher educational programs.

9.2 Policy Measures

This section presents several policy measures that can be undertaken in Albania and other developing countries in order to intensify the qualitative effects of FDI in the host territory.

9.2.1 Policy Measures with Respect to the Clothing Industry

Obtaining industrial upgrading in the clothing industry requires a set of policy measures that include but are not limited to:

- Creation of a stable local environment for doing business in Albania by removing (fiscal barriers, custom burdensome procedures, uninterrupted power supply, etc.) that will reduce any uncertainties and encourage foreign investors to locate production in the host territory including complex functions like design, marketing, etc. The external/international environment of doing business also affects the stability in the local environment. Despite of the fluctuations in the external environment, which cannot be controlled by local authorities, the long term location advantages play a key role in retaining attractiveness for foreign investors. Creation of economic zones would permit both foreign and local enterprises operating in the clothing and in the manufacturing industry to grasp the benefits of having in the same area manufacturers, suppliers of raw materials, or subcontractors. This could benefit Albania as it would facilitate connection between suppliers and producers, a problem encountered in Albania also in other industries in which local suppliers are available. The economic zones will serve as a good example to enterprises in the clothing industry and to those in other industries on how to realize a full production cycle. This can be achieved by creating the first economic zone in the region of Durres or Vlora where major ports are located so to ensure even a quicker delivery time of finished articles. China is one of key players in the world economy that has created successful economic zones where enterprises rely on each other to realize full production cycle. In these economic zones, some enterprises serve as suppliers, others as producers, and the remaining as distributors. Based on the example of economic zones in China, the Government of Albania needs to create the first economic zone in Albania close to a major port. Special economic zones are zones that have different trade laws and taxes from the rest of the country. They have one or more objective: (i) attracting FDI, (ii) acting as a laboratory for trying new policies or (iii) creating employment. In terms of integration into GVCs, economic zones shorten bureaucratic delays and provide good infrastructure better networking opportunities, freedom from import duties, and absence of a number of taxes.

- Policies need to focus not only in attracting new FDI but also in increasing integration of existing FDI in the host territory. Integration in the host territory of existing FDI can be achieved by looking for possible local suppliers, train them, and introduce them to clothing manufactures both foreign and domestic. Up to now this policy has been applied on suppliers of products like hand-made lace used in manufacturing of delicate products. Another policy to promote local suppliers is to impose for every foreign enterprise a minimum local requirement of raw materials on the goods manufactured in Albania. In order to impose these requirements the Government of Albania needs to strengthen its location advantages by making locally accessible raw materials that meet international standards so that the share of imports on these kinds of articles is reduced.

- Institutions in Albania can facilitate a higher degree of integration of foreign investors in the host territory, especially by allocating more responsibilities to the Albanian Investment and Development Agency (AIDA). These responsibilities would not be limited only at attracting FDI in the country but also in retaining investment and expanding the activity of existing foreign investments in Albania. As such, a new department that will monitor and get continuous feedback on the activity of foreign enterprises in general and of clothing manufacturing enterprises in particular needs to become functional within the agency. In order to support foreign investors located in all regions of Albania, it is necessary to open regional branches of AIDA that will facilitate the doing business of foreign enterprises in all regions of Albania. These responsibilities can be allocated to AIDA by approving al the required legal framework as a Decision of the Council of Ministers.

- In order to participate in GVCs, it is also crucial to start building a national innovation system. This system is a stepping stone with regard to further integration into GVCs. Developing countries should create institutions that assist in the technology transfer between all stakeholders in the GVCs. Technological capabilities are more important since increased scientific content is valued more in fields such as electrical and chemical engineering, telecommunications, electronics and biotechnology. It is now more important to provide advanced training in science. Firms should have independent capabilities to develop technologies. A GVC perspective needs to be adopted when thinking about the innovation system. Foreign buyers need to be integrated in local innovation systems, integrating them in joint R&D projects. By taking the upgrading trajectory of firms into account, priority areas for direct funding of research should be identified. Populating the innovation system with research and service providers that are important to the value chains will enable upgrading in those areas.

9.2.2 Policy Measures with Respect to Knowledge Transfer

In order to acquire more knowledge from the presence of clothing manufacturing subsidiaries the Government of Albania can initiate several policies.

Firstly, promotion of fiscal incentives for acquisition of new investments in physical capital in the form of modern machines that will enable local employees to learn new processes and operations. In order to benefit from the investment made the clothing manufacturing enterprise will hire and train a number of employees. For example, Latin American countries have launched in cooperation with financial institutions new financing programs that increase the level of credit in the economy.

A second policy option is by undertaking national campaigns that would promote training of employees in the clothing industry. A national vocational program that supports clothing manufacturing enterprises to organize trainings for local employees that span beyond the minimal skills required for assembly services needs to be created. The Government of Albania can intervene by making the vocational training program compulsory for clothing manufacturing enterprises at all levels. Training of local employees can be facilitated through creation of national training centers based on a triple helix model that involves: (i) educational institutions, (ii) clothing manufacturers, and (iii) government institutions. Malaysia is one of the countries that has undertaken nation spread programs to increase the skills of the labor force in the clothing industry through intensive programs run by international experts. A similar training program is undertaken also in the electronic industry. For instance, specialized research institutions can provide knowledge and resources at the technical level such as machinery maintenance, as well as at the human level such as training and capacity building provision to local firms. Capacity building can be in areas such as quality control, safety measures etc. One of the problems in developing countries where such institutions exist, is that they are underfunded and not working properly in assisting the local firms. Sharing knowledge is very important for the integration in GVCs. That's why specialized institutions are crucial to facilitate the interaction between all stakeholders such as domestic firms, foreign firms, financial intermediaries etc.

A third policy option can occur by developing backward linkages with the local economy through financial incentives in the form of tax reductions, subsidies to finance modern machines targeting of producers that can serve as suppliers of raw materials for clothing manufacturers in the country. The presence of more sophisticated machines can bring additional knowledge that will strengthen technical capacities of local employees working for local suppliers and ensuring production of raw materials like card boxes and plastic bags to meet the standards required by international customers in the clothing industry.

A fourth policy option refers to launch a series of trainings to strengthen technical abilities of local management in running clothing manufacturing subsidiaries. These trainings can run in coordination with educational institutions that will provide these trainings. Establishment of regional training centers that would identify the needs of employees in the clothing industry and organize appropriate training programs can spread the required skills throughout the country. This way, the trainings will address not only the needs of employees but also will be in line with any developments occurring in clothing manufacturing enterprises.

9.2.3   Policy Measures with Respect to Subsidiary Upgrading

In order to facilitate upgrading of clothing manufacturing subsidiaries in the Albanian economy a number of policy measures can be undertaken which include but are not limited to:

- Reduction in the time required for processing of imports of raw materials and exports of finished goods by eliminating redundant custom procedures and by reducing paper work through digitalization of services in all custom points in Albania. This would save the time of the high and middle management that are daily engaged on these issues allowing them to dedicate it to changes in the operational activity of the subsidiary like engaging in more complex processes or manufacturing of more complex articles.
- Commitment of the government to strengthen the power supply network in areas in which industrial production occurs throughout Albania. This would save clothing manufacturing subsidiaries financial resources required to purchase and maintain power supply generators. These resources can be invested in purchasing technologically more advanced machines or in training programs for employees needed to perform in the host territory more advanced operations.
- Opening of branches of international design schools in Albania in order to educate quality designers in the country, increasing this way the possibility for partial/full transferring of the design function in Albania. This can be achieved through twinning projects between international schools and local institutions already engaged in provision of study programs on fashion design. Through these projects, local firms can acquire more knowledge in the field that can be of use to clothing manufacturing enterprises for the design of goods they manufacture. This policy is beneficial also for local clothing manufacturing enterprises that would like to expand beyond assembly provision of services to their customers.
- Promotion of the clothing industry in Albania by increasing its capability of participating in more segments of the value chain of the clothing industry (fabric, design, etc.) and ensuring the local clothing industry moves up in the value chain. By following such policy, Albania may rely less on low wages to attract foreign investors that can afford to make large investments on advanced machines and initiate processes like design and research and development on which local enterprises in Albania do not possess adequate knowledge.

9.3   Additional Areas for Research to Facilitate Integration into GVCs

Additional research is required to maximize qualitative effects in the host territory. These areas can include but are not limited to:

- Identification of areas that hinder local firms to become suppliers of clothing manufacturing subsidiaries in Albania together with the actions that are necessary to fill in the gaps. Among the proposed actions can be identification of current manufacturers of potential raw materials in Albania, and launching of a national dialogue between clothing manufacturers and suppliers. Creation of a base of local suppliers that meet the needs of clothing manufactures may serve as a stepping stone in the transferring more processes and functions in Albania.
- Investigations of the electrical distribution network in order to identify the mistakes made in the implementation of previous reforms that were unsuccessful to reduce or eliminate interruptions in the power supply and to propose concrete actions that need to be taken in order to provide short and long-term solutions.
- Investigation in the quality of products manufactured in Albania with regard to those manufactured in Asian Countries and to identify improvements needed to be made so that more production is transferred from Asia to Albania. Moving up in the value chain of the clothing industry especially on how to revitalize the manufacturing of various fabrics that ceased to be functional in the 1990s.
- Undertaking due diligence and feasibility studies on the creation of the economic zones in Albania including possible scenarios on the impact it might have in the development on the clothing industry and local economy.

Glossary

Case study An empirical study that investigates a contemporary phenomenon within its real life context to understand complex social phenomena.

Cluster Refers to groups of similar and related firms in a defined geographic area that share common markets, technologies, needs for employee skills, and often linked by buyer-seller relationships.

Economic development Qualitative change and restructuring in a country's economy in connection with technological and social progress reflecting an increase in the economic productivity and average material wellbeing of a country's population. Economic development is closely linked with economic growth in the long run.

Embeddedness Integration of a subsidiary in the host territory through the linkages and interaction with local suppliers, generated employment, and the level of investments made in the host territory.

Endogenous development Economic development based on internal determinants rather than on external ones. It highlights that investments in human capital, innovation, and knowledge trigger economic progress while long-term economic growth is sustained by capable institutions, adequate regional development, and appropriate policymaking.

Exogenous development Development of a country based on imported technology, capital, and human resources. In this approach to development, external agencies/actors participate in the process of development.

External validity Generalizing the findings of a case study to greater groups than those analyzed in a case study. Generalization is not automatic, as it should be based on theory.

Fast fashion Quick replenishment of clothing, which in turn allows the retailer to offer a broad variety of fashion clothes without holding a large inventory.

Focused interview An interview designed for a short period and follows a set of prepared questions.

© Springer Nature Switzerland AG 2020

J. Kacani, *A Data-Centric Approach to Breaking the FDI Trap Through Integration in Global Value Chains*, Lecture Notes on Data Engineering and Communications Technologies 50, https://doi.org/10.1007/978-3-030-43189-1

Foreign direct investment The net inflows of investment to acquire a lasting management interest (10% or more of voting stock) in an enterprise operating in an economy other than that of the investor, often referred to as the host territory.

Knowledge It refers to facts, information, and skills acquired through experience or education. Theoretical and practical understanding of the subject. It ranges from abstract ideas such as scientific formulae to eminently practical ones such as traffic circle.

Long term sustainable The use of economic resources like physical, human capital, growth and technology more productively and efficiently. In particular, it redirects attention away from capital accumulation, plant capacity, and acquisition of equipment to immaterial resources like innovation, human capital, and knowledge.

Industrial upgrading at a subsidiary level It occurs when the subsidiary gets more embedded within the host economy and realizes progressively more complex activities.

Innovation Creation and diffusion of new ways of doing things.

Internal validity Establishing causal relationships among the variables taken into account in a case study research. It includes a great deal of inference derived from relating the variables with each other.

Multinational enterprise Companies or other entities established in more than one country and so linked that they may co-ordinate their operations in various ways. While one or more of these entities may be able to exercise a significant influence over the activities of others, their degree of autonomy within the enterprise may vary widely from one multinational enterprise to another.

Open ended interview A casual interview that does not follow a fixed set of questions.

Reliability Refers to demonstrating that the steps of a case study such as data collection procedures can be repeated.

Qualitative effects of FDI Refers to the impact of foreign direct investments on knowledge transfer, creation of a base of local suppliers for intermediary goods, creation of a qualified labor force, and the evolution of the subsidiary.

Quantitative effects of FDI It refers to the impact of foreign direct investment of the job creation, GDP, physical stock of capital, tax revenues of the host territory, etc.

Subsidiary An enterprise that is controlled by head office of the multinational enterprise it belongs to.

Technology It is useful economic knowledge that comes from new consumers' goods, the new methods of production or transportation, the new markets, the new forces of industrial organization that capitalist enterprise creates.

Transfer pricing The price that is assumed to have been charged by one part of a company for products and services it provides to another part of the same company, so to calculate each division's profit and loss separately.

Value chain The full range of activities that are required to bring a product or service from conception, through the different phases of production (involving a combination of physical transformation and the input of various producer services), delivery to final consumers.